THEY SAID IT COULDN'T BE DONE

THEY SAID

E. P. Dutton New York

IT COULDN'T BE DONE

Ross R. Olney

Library of Congress Cataloging in Publication Data

Olney, Ross Robert, date They said it couldn't be done.
Includes index.

SUMMARY: Describes 10 feats of engineering that were believed to be impossible including the construction of the Brooklyn Bridge, the Empire State Building, the Hoover Dam, and the landing on the moon.
1. Civil engineering—United States—Juvenile literature.
2. Space flight to the moon—Juvenile literature.
[1. Civil engineering. 2. Space flight to the moon] I. Title.
TA149.046 1979 624'.0973 78-12405 ISBN: 0-525-41060-0

Published in the United States by E. P. Dutton, a Division of Sequoia-Elsevier Publishing Company, Inc., New York

Published simultaneously in Canada by Clarke, Irwin & Company Limited, Toronto and Vancouver

Editor: Ann Troy Designer: Patricia Lowy

Printed in the U.S.A. First Edition
10 9 8 7 6 5 4 3 2 1

The author and publisher gratefully acknowledge permission to quote from *The Empire State Building* by Theodore James, Jr., on page 58. Reprinted by permission of Harper & Row Publishers, Inc.

Jacket photos credits:

Front, top, left to right: Mount Rushmore National Monument, National Park Service; The Brooklyn Bridge, Museum of the City of New York; The Empire State Building, Museum of the City of New York; The Gateway Arch, Bi-State Development Agency. *Bottom:* The Moon Landing, NASA.

Back, top, left to right: Hoover Dam, Las Vegas News Bureau; The Golden Gate Bridge, Bethlehem Steel Corporation; The Astrodome, Houston Sports Association; The Indianapolis Speedway, Indianapolis Motor Speedway. *Bottom:* The Holland Tunnel, The Port Authority of N.Y. & N.J.

Title page spread photo credits: Left page: Same as back of jacket. *Right page:* Same as front of jacket.

Acknowledgments

The author would like to thank the following persons and organizations for photos, advice, and anecdotes:

Don Beale, Las Vegas News Bureau
Al Bloemker, Indianapolis Motor Speedway
Frank G. M. Corbin, Economic Development Administration, City of New York
Paul Darst, Houston Sports Association
Gladys Hanson, San Francisco Public Library
Ed Hoffman, South Dakota Division of Tourism
Lyndon B. Johnson Space Center, NASA
Kurt L. Malmgren, Bethlehem Steel Corporation
J. Dianne Rains, Bi-State (Illinois and Missouri) Development Agency
Kenneth Vick, U.S. Department of the Interior, Boulder Canyon Project
Cindy Westbrook, San Francisco Visitors & Convention Bureau
Harvey D. Wickware, U.S. Department of the Interior, Mount Rushmore National Monument
Thomas C. Young, Port Authority of New York
James R. Youse, U.S. Department of the Interior, Jefferson National Expansion Monument

and especially Lucia Greene of E. P. Dutton, who located some photographs I thought were impossible to find. Without them and material provided by all these others, the book would be only half-finished.

To my longtime friends
Don, Mary Lou, and Shawn Zents

Contents

HIDES SKINS&C
HIDES

THE BROOKLYN BRIDGE

The Brooklyn Bridge about 1890, looking from Manhattan to Brooklyn (Bethlehem Steel Corporation)

The winter of 1866–67 was one of the worst in history in New York City. Snow blanketed the city for most of the winter, cold winds blew around the buildings, and great ice floes choked the East River between Brooklyn and Manhattan. Very often ferryboats were unable to cross, so long lines of commuters trying to reach Manhattan or return home to Brooklyn were stalled at the riverbank.

Tempers grew hotter as the temperature grew colder.

It was a time to joke about the poor people who lived in Brooklyn. Everybody in the country knew that travelers could reach Manhattan from Albany quicker than they could from the city just across the East River.

For in 1866 there was no Brooklyn Bridge to sell to the yokels, and the ice-jammed river was a formidable barrier. Engineer John A. Roebling, though he had no idea of the tragic cost to himself and his family, had suggested that one of the newfangled wire-hung bridges be built between the two cities. Roebling had perfected a cable-weaving technique that was supposed to make such bridges safer. He had just completed yet another fine wire bridge across the Ohio River. He was building bridges everywhere, and even people in the Midwest were joking about New York's backwardness in rejecting the perfectly good idea of a Roebling bridge over the East River.

John A. Roebling was the leading bridge builder of the day. He was founder and president of the John A. Roebling Wire Rope Company and an expert on suspension bridges.

Not that suspension bridges in general had that great a reputation. Brooklyn residents may have been slow to accept new ideas, but they weren't dumb. They could read about Dryburgh Abbey Bridge across the Tweed River in Scotland and several great bridges in England, all of them soon wrecked by wind action. Some of them had been wrecked *twice*.

And it didn't take a genius in engineering to research the world-famous 580-foot Menai Strait Bridge in Wales. It was subject to alarming aerodynamic undulations, sometimes twisting and bouncing in *16-foot waves*. Wind action broke it down in 1826 . . . and again in 1836 . . . and again in 1839. The Nassau Bridge over the Lahn

The Brooklyn Bridge today, looking from Brooklyn to Manhattan. Lower Manhattan has changed drastically since the bridge was built. (Museum of the City of New York)

River in Germany broke down, and so did the Roche-Bernard Bridge in France. The latter one dropped a railroad train into the river.

All of these were wire suspension bridges.

The Wheeling Bridge, a suspension bridge over the Ohio River, "lunged like a ship in a storm," according to eyewitnesses, and plunged into the river after completely *overturning.* The Lewiston–Queenston suspension bridge over the Niagara River, a then record span 1,043 feet long, broke in a gale in 1855.

The disasters went on and on, so the people of Brooklyn had always just smiled when others asked them why they didn't want a nice bridge across the river instead of fighting the ferryboats. They knew about Roebling's plans for a wire suspension bridge. Let him build it someplace else, a place where the people were looking for some *action.*

But Roebling insisted he had perfected a new technique, and by the cold winter of 1867 the residents of Brooklyn were willing to take almost any chance at all if it meant a quicker trip across to Manhattan, where thousands of them worked. By that year the demand for a bridge had risen to a clamor, and an initial charter was granted. Roebling, to his own great pleasure and satisfaction, was appointed chief engineer. Brooklyn's transportation problems were coming to an end.

But John Roebling's problems were just beginning. Before the great bridge would span the river, Roebling's son would be paralyzed forever . . . and Roebling himself would be dead. The Roeblings paid heavily for the beautiful monument they left behind over the East River in New York City.

In his Ohio River bridge, Roebling had added something new to suspension bridges—cable stays running diagonally upward from the stonework of the towers to points of the cables in both the main span and side spans. One of the most distinctive features of the Brooklyn Bridge is the system of inclined stays radiating downward from the towers to the floor of the spans.

With these extra stays, the bridges are guaranteed to stand for generations. Finally the beleaguered people of Brooklyn believed what Roebling said, and on the second day of January 1870, construction began.

But not under the guidance of master bridge builder John A. Roebling, who had dreamed of the great bridge over the East River, connecting Brooklyn with Manhattan.

During the summer of 1869, Roebling, who had designed the details of the entire project, was on the Fulton Street ferryboat surveying the exact location of the Brooklyn tower of the bridge. The timber-loaded boat thrust into the Brooklyn slip and shifted some wood on which he stood. His foot was crushed.

Roebling contracted tetanus, but even then he might have been saved. He fought his doctors to the end, refusing to take their medications and refusing even to stay in bed until his final days. He had work to do on his great bridge, he insisted, and no doctors were going to keep him from it. His foot would heal, he assured them.

It did not, and sixteen days after the accident, John A. Roebling died.

Macabre stories about the bridge claiming a life in order to be a safe bridge began to circulate.

Roebling, the master bridge builder, was dead. But the project did not stop, nor did this end the tragedies of the Roebling family. Roebling's son, Colonel Washington A. Roebling, thirty-two-year-old civil engineer himself, was appointed to take over his father's bridge. He gladly accepted the chance to prove his father's theories and lived with the bridge for most of the rest of his life.

Roebling built the bridge many times beyond the

After construction began in 1870, dignitaries enjoyed having their pictures taken on the site . . . but not too high up. (Museum of the City of New York)

safety limits of the day, and it was a good thing. Today the Brooklyn Bridge carries traffic undreamed of in Roebling's day, loads far in excess of what it was expected to carry in the 1800s. Every piece of anchorage for the huge cables, for example, was tested "fourfold," according to the engineer's reports of the time. Testing was done "by means of specimen pieces [of the anchorage] under the enormous power of the hydraulic press to the breaking point, a wide margin being always required above the highest possible strain that it is estimated can ever come upon it."

Roebling was taking no chances with the names of himself and his father.

The anchorages far inland and the towers were gradually completed, but not without problems. The towers required caisson work, and caisson work under air pressure is always dangerous. It was especially dangerous during the years of construction when people knew less than they do today about the hazards of breathing compressed air.

The two towers needed to be anchored to the bedrock of the earth under the riverbed. First, huge chambers called caissons were sunk to the bottom; then, with compressed air, the water was pumped out. As the tower work began on top of the chambers, workmen entered the chambers through air locks, to dig away at the river bottom. A constant air pressure was maintained in each caisson to hold back the river water.

Men would work for long periods of time in the increased air pressure of the caissons far beneath the ever-climbing towers. They would dig away at the mud and silt, reaching for bedrock. Aboveground the people of New York gazed in wonder at the slender towers reaching higher and higher into the sky. They clucked and worried and asked themselves if such towers could actually hold the weight of the great bridge. Underwater, on the river's deepening bottom, tunnel-digging men called "sandhogs" worked away.

Much is now known of the terrible pressure disease called bends or caisson disease. Like a carbonated soft drink capped in a bottle, the blood of a person under pressure stores nitrogen in solution. When the person returns to the surface, as in a soft drink with the bottle cap suddenly released, the bubbles of nitrogen come out of solution and try to reach the outside air.

They collect in the joints and cause severe pain and in some cases permanent injury or even death. It is now known that the only treatment for this terrible condition that twists a person's body in agony is to bring the body back to high pressure. This returns the nitrogen to solution in the blood, and the condition is stabilized. Then, by very slowly reducing the pressure to normal, according to complicated charts that have been worked out, the nitrogen is released without injury. Deep divers must go through decompression stops on their way back to the surface, or else they must be immediately recompressed in a recompression chamber and then decompressed slowly and by the charts.

Many workmen in the deep caissons at the bridge site suffered the bends. One was the chief engineer, Colonel Washington A. Roebling, himself. Roebling was not one to sit in his office and supervise construction by remote control. He crawled up the towers and descended to the river's bed to personally watch over the job. He was hit with the bends several times, but one final devastating attack after a long period in a caisson ruined his health. He knew he was staying down too long, and he probably knew he was coming back to he surface too quickly, but he had *work* to do. The bridge would not wait.

He was paralyzed for life because of the accident. From that day on he sat at a window of his nearby apartment, watching the bridge through field glasses, watching the bridge that had claimed his father's life and his own health.

But *not* helplessly. Washington Roebling was a dynamic man and a superb engineer, a man who wouldn't quit no matter what the odds against him.

Unable to speak and barely able to move, he devised a hand-tapping code with his remarkable wife, Emily. Roebling would study the bridge through his glasses, then send instructions to his supervisor and engineers by tapping out messages on his wife's hand. She would pass the instructions on, and the bridge continued to grow.

There were other accidents at the construction site, including a terrible fire deep inside the caisson on the Brooklyn side, but work went on. To extinguish the caisson fire, firemen merely flooded it with water, but work was slowed during the caisson repair.

Both huge caissons are still there, deep under the bridge towers. They were finally filled with concrete under hydraulic pressure to unite with the bedrock, giving each tower a massive and immovable foundation. Then the towers were finished. The Brooklyn tower was finished in May 1875. The New York tower, the more difficult one because of the great depth to which men had to

Finally the cables were started. The entire roadway was to hang on these thin cables. The steel anchor plates in the foreground now extend deep into cement at the anchorages. (Museum of the City of New York)

dig through sand and gravel to reach a solid footing in bedrock, was finally finished in July 1876, more than a year later.

But the great towers were there only to support the weight of the wires that would rest over them in huge cradles. It was to be the wires that would hold up the bridge. Each wire cable was to be made of 19 strands according to the Roebling Wire Rope Company, the world's experts in wire spinning. Each strand would be made up of 278 separate No. 8 wires. Each wire would, according to the engineers of the day, be "eternal in length." There would be no beginning or end to break apart, but instead a single long length traveling from anchorage to anchorage and up over the towers.

There would be no twist in the wires, such as in a rope. They would lay parallel to and against each other, bound into a circle, strand by strand, then all of the strands joined together into four mighty cables each *fifteen inches* thick.

These four would hold up the roadway.

New Yorkers wondered how the wire bridge could be started between the two beautiful towers. But the beginning of the process was uncomplicated. A boat simply carried a three-quarter inch wire rope across the river after river traffic had been temporarily stopped. The coil was pulled up and over one of the towers by another rope. A workman far up on top of the other tower then lifted the wire rope up, and the connection was made. That single strand, a spider web high in the sky and almost invisible to spectators, was the beginning. A second wire rope was installed with the first one, the two were attached at each end around great rotating drums, and the endless cable was completed. By this cable, the first wire-rope spinning would be done.

But not before one of the workmen, E. F. Farrington, under the envious gaze of the injured Washington Roebling, hooked up a bosun's chair and shot out on a dizzying ride between the two towers, high over the river. A million people watched from both shores as Farrington sped down the sag from the Brooklyn tower and was pulled back up to the top of the New York tower.

The great river had been officially bridged by a man. Bands played, cannons roared, and steam whistles screamed from vessels on the river far below. It had been six and one-half years since construction began. It would be seven more long years before the bridge would finally open to traffic.

As cable was strung between the towers and to the anchorages, problems with weather and heat expansion plagued the builders. If the sun was shining on one section and not on another, measurements would change drastically. Deflection of wires varied one-third of an inch for each degree of temperature. The wind played havoc with the wires.

"In short," said one writer of the day, "the ponderous bridge, while neither small nor agile, has a trick in common with the minute and lively insect which when you put your finger on him isn't there."

But finally the cables were in place, the suspender cables were bound, and the roadway

The four fifteen-inch-thick cables are shown in detail as the roadway is being hung. Two visitors are enjoying a magnificent view of the project from high on the workers' catwalk. (Museum of the City of New York)

was started. Huge expansion joints were built in so that the bridge roadway could actually expand or contract during temperature changes without collapsing.

People watched but could still barely believe that the weblike structure of wires would hold up the massive roadway. Enormous weight was hanging in thin air. Many said they would never cross the bridge under any circumstance. But for a few daredevils, it was proof of bravery to walk up from the anchorage, down the long sag from the first tower, up the hill to the second, and down to the other anchorage on a thin, slatted, one-man-wide catwalk built by workmen to service the cable-spinning machine high above the roadway.

Construction sites were not closed then as they normally are now, and construction men didn't care if you were fool enough to risk your life for nothing, doing what they had to do to earn their pay. Just as long as you stayed out of their way.

Finally, on May 24, 1883, the festive day arrived. Sitting alone at his window, his field glasses trained on the great celebration on his bridge, Roebling watched as top-hatted dignitaries strolled under the high arches. It is possible, insist some New Yorkers, that even today, as the winds play against the cables in the night, you can hear the echoes of the opening-day bands. They played and played as President Chester A. Arthur and his cabinet stared in awe at the massive towers and the framework of delicate wires.

The governor of New York was there, and the mayor of New York City, and military exhibits and throngs of ordinary people stretching all the way up Broadway. Spectators by the tens of thousands jammed lower Manhattan.

The great bridge was *open!*

On the river below, the U.S.S. *Tennessee,* flagship of the Atlantic Squadron, belched forth with repeated cannon fire. Several other huge navy ships and thousands of smaller boats steamed back and forth under the bridge, shooting and whistling. On the bridge a great calliope shrieked out "America" over and over again.

Roebling could hear. He watched and smiled. When others had doubted, he had known the Brooklyn Bridge would be built successfully. And it was.

9
8
5

THE INDIANAPOLIS SPEEDWAY

Joe Jagersberger (No. 8) hurtles along moments before his crash in 1911. Others are (No. 5) Louis Disbrow in a Pope Hartford and (No. 9) Will Jones in a Case. (Indianapolis Motor Speedway)

The big Case automobile of race driver Joe Jagersberger hurtled, smoking and snorting, down the back straightaway at more than seventy miles per hour. It careened through turns three and four and started down the main stretch of the brand-new raceway in full view of the thousands of fans.

Not long before, the very same area had been a cow pasture on a deserted farm near Indianapolis, Indiana. But Indianapolis was a powerful city in the new "automobile" business, so the racetrack idea had seemed a good one to Carl G. Fisher.

But not at that moment to Jagersberger. He looked in horror at his tie rod, a unit connecting the front wheels of his racer. The rod had suddenly snapped. The front wheels of the racer splayed apart; one pointed one way and one the other. The racer was completely out of control!

Jagersberger couldn't even *aim* the hurtling racer. All he could do was put on the brakes and hope that nobody slammed him from behind.

Engineer Ray Harroun, riding alone in his Marmon Wasp, finally won the first 500-mile race at Indianapolis Motor Speedway. (Indianapolis Motor Speedway)

Mechanic C. L. Anderson, Jagersberger's companion in the Case, could see that a good swift kick at the right wheel would snap it over to the left, and the car would then head to the inside wall. Anderson knew that the wheel-kicking job was his.

He waited until the car began to slow . . . waited . . . waited . . . and finally he *jumped,* intending to catch up to the wheel and give it a kick. What he forgot was that the car had been going over seventy miles per hour. So thirty miles per hour felt almost like standing still. The mechanic went sprawling and rolling down the track beside the car.

Screaming onto the scene in his Westcott racer came driver Harry Knight. Immediately in front of the wheels of the Westcott was poor Anderson, stunned and unable to scramble. The code of auto racing then as now is that if you are going to crash, crash alone and take no other man with you if you can help it. Knight followed the code to the letter, jerking his steering wheel left and missing Anderson by scant inches.

But in doing so, he skidded broadside, sideswiping Jagersberger's Case and finally slamming into Herbert Lytle's Apperson, then in the pits for service. Jagersberger hit the wall hard. All three cars were destroyed, and Knight was seriously injured. So was his mechanic, John Glover. So were Jagersberger and Anderson.

This was only one of several incidents in the inaugural 500-mile Indianapolis auto race finally won by engineer Ray Harroun, driving a Marmon Wasp. Louis Disbrow and "Terrible" Teddy Tetzlaff had also collided on the front straightaway. Arthur Greiner had spun and slammed into the southwest wall. Mechanic Sam Dickson had been killed in that crash.

It all happened on May 30, 1911, when auto racing was young and even more untamed than it is today. That is why they all told Fisher that he was foolish to invest his hard-earned money in something as childish as motor racing. Sure, the new automobile was fun, but it was only a toy. Yes, there were some races around, but nobody would take seriously a several-hour race on a giant track on a farm outside town. Nobody.

Two years before, Fisher had envisioned the vast track. An Indianapolis resident, he had built his fortune in the auto business. He joined James Allison, Arthur C. Newby, and Frank H. Wheeler. The group optioned the abandoned Pressley farm near Indianapolis. The asking price, though high at $72,000 for the 320 acres, was too much to resist.

As rumors spread through Indianapolis and the automobile world about a new Indiana Motor Parkway Grounds, the official papers were filed. The new company was called Indianapolis Motor Speedway Company. Carl Fisher was president.

Today it would be difficult to put a value on the Indianapolis Motor Speedway's track, grandstands, grounds . . . and reputation. Millions of dollars are paid every year to the racers in the big race. "Indy" attracts spectators from around the world. The calm midwestern city of Indianapolis, like a great ocean liner steered suddenly into a hurricane, comes to fast and vivid life during May. Everybody in the automobile world and thousands and thousands of people with no connection at all to cars stream to Indianapolis in May. They attend meetings and parties and qualification runs, and as the end of the month approaches, the pace becomes more and more frantic. In the last few days there isn't enough time to do everything. Then comes the big race.

But in 1909 there was a long step between owning an abandoned farm outside of town and owning the world's greatest motor racetrack. Sure, Indianapolis was an important town in the new industry, but the racing being done then was on

Carl G. Fisher, founder and first president of Indianapolis Motor Speedway, speeds around the track while it is still under construction. His passengers are reporters from local newspapers. (Indianapolis Motor Speedway)

shorter tracks. Racing already had some daredevils like David Bruce–Brown, Ralph DePalma, Howdy Wilcox, and Bob Burman, but the vast five-mile track envisioned by Fisher and his pals was a great gamble. Fisher was certain a long race on a huge track would be popular, but most people and even some racing drivers questioned whether fans would attend and sit through such a race.

The famous 500-mile races were not the first at the Indianapolis Speedway. But the first events almost ended the whole project for Fisher and his associates before the 500-mile races began. The plan was to introduce the great track to the public with a series of races over several months, including motorcycles, hot-air balloons, and finally the fastest race cars in the world in a 300-mile grand finale. Then later, Fisher hoped, could come the 500-mile races.

Workmen struggled to get the track ready for opening day. The decision had been made to make the track a smaller two and one-half mile course with two long straightaways and two short ones, to form a rectangle with rounded corners. The idea was that at some time in the future another grandstand or two might be needed, so space should be left on the property outside the track for this possibility.

The modern Indianapolis Motor Speedway now

seats nearly *one-third of a million people* with room for many more in the infield; so the decision was a good one.

At first 100 workmen toiled on the old Pressley farm; then 350 more were added as time grew shorter. Fisher brought in 300 mules, 150 road scrapers, 4 six-ton rollers, and 3 ten-ton rollers. Construction of the racing surface was the difficult part. The preliminary grading came first; then a two-inch layer of creek gravel was spread over the entire two and one-half mile surface. This was rolled and compacted by the ten-ton rollers. Workmen followed along, adding two inches of crushed limestone with two gallons of taroid, a paving substance, per square yard.

The third step called for a coating of one-half inch of crushed stone chips, with a top dressing of stone dust worked into the crevices by the six-ton rollers. Then a final coating of taroid was applied.

With carpenters completing the two grandstands (long since torn down for new modern stands almost all the way around), the track surface began to look smooth and *very* fast.

But it turned out to be a *disaster*.

A hot-air balloon race, including Fisher's own *Indiana*, was the first event ever held at the nearly completed Indianapolis Motor Speedway. Three thousand people paid to come in, and 40,000 more watched from outside as the balloons rose majestically over Indianapolis. Balloonist John Berry's *University City* was the winner, covering 382 miles, all the way to Fort Payne, Alabama.

Frantically the smoothing work continued on the track for the first races scheduled for the following August, 1909. These were to be a series of motorcycle races with the champion riders of the day entered. Though famous riders did appear (J. F. Torney, John Merz, H. R. Bretney, and E. G. "Cannonball" Baker were some), the event was termed a complete fiasco by the newspapers. The track was not ready, and many accidents resulted. The news media blasted the new track, and it appeared that the speedway dream was over before the first automobile ever tested it. Without spectators, no motor race can succeed.

But Fisher and his associates were men of vision who were certain that the track could succeed. The first auto races were less than a week away, so they desperately worked to revive public interest in the track. They took out advertising, they spread handbills and posters, they gave tours of the track, and they planted items about famous drivers in the newspapers. They tried everything they could think of to keep the track alive in people's thoughts.

On the morning of the first day of the three-day event, as dawn was breaking in the east and workmen were still toiling to get a final layer of oil on the main straightaway to contain the dust, Fisher waited. He had done all he could. He and his pals watched as the early morning interurban cars began to roll in. The grandstands were ready and waiting, the parking facilities arranged. Free parking included 3,000 hitching posts and room for 10,000 cars. General admission was fifty cents or one dollar, depending on where you wanted to sit. Then there were several thousand box seats for an additional dollar fifty.

Sunlight crept down to fill the shadows in the empty boxes. Gradually the workmen finished and left the track. All was ready. Then the reports began to filter in to Fisher who was in the main press box.

Long motorcades were beginning to arrive at the gates from most of the big cities in the Midwest. In one of them, in fact, was a young fifteen-year-old by the name of Tommy Milton. Milton was destined to become the first two-time winner of the annual 500-mile race (in 1921 and 1923).

Meanwhile, trains and interurbans loaded with

fans puffed into the station by the main gate. Long before noon, the two grandstands were jammed to their 15,000-fan capacity. It seemed that Fisher had won his gamble once and for all.

But by the final race of that first day, the racers were struggling through a blinding shower of oil and dust and sharp-edged stones. The track surface was breaking up so quickly that officials considered stopping the races. Driver injuries and worse were piling up.

The Knox racer of William Bourque hit a chuckhole in the failing track surface, flipped upside down, and both Bourque and his mechanic, Harry Holcomb, were killed. Bob Burman finally won the last event of the day as officials were pondering what to do. They considered calling off the next day's races.

"No," said Fisher. "Give us a chance. We can repair the track and have it ready," he promised the sanctioning American Automobile Association.

The second-day races were held before an even larger crowd, and the only injury of the day was to the legendary Barney Oldfield. Oldfield received a bad cut on his arm while shielding his face from the hood of his racer. A fire in the engine had burned the leather straps holding the hood on as Oldfield typically tried to speed the burning car back to the pits for help. The hood had blown off and onto the driver.

But once again the track surface failed and had to be patched.

The final day of the first auto racing event ever at Indianapolis could hardly have been worse. Still, in the grandstands and infield, there were more than 35,000 people, by far the largest crowd of the grand opening.

The three preliminary races of the final day were won by Eddie Hearne, Barney Oldfield, and Ralph DePalma without incident. Then came the 300-mile Wheeler–Schebler Trophy Race. Seventeen of the world's most powerful racing cars were in the starting lineup. The track surface was to be subjected to its greatest test.

Minor accidents began as the stone surface failed. Drivers couldn't see each other in the dust and haze. While this problem grew, another developed. In spite of repeated warnings, fans began to crowd along the outside walls of the turns. All drivers and many fans were in great danger. More than one driver sighed with relief as his racer failed, forcing him out of the contest. Many complained that tragedy was almost certain if the race continued. But racing men race, so in spite of the danger, these men raced on. There were more skids and spins as the dust increased. Drivers were being injured by flying rocks. Spectators continued to crowd the walls.

Then it happened. The National of Charles Merz skidded out of control and climbed over the top of the wall. It soared through the air more than 100 feet and finally plowed through a cluster of spectators. Then it bounced across a creek and landed upside down in the soft mud. Merz crawled from under the car, unaided.

But his mechanic, Claude Kellum, who had been in the car for only a few laps to relieve the regular mechanic, was killed instantly. Also killed were Indianapolis resident James West and Franklin, Indiana, resident Homer Jolliff. Several others were injured.

As the accidents continued, officials on the starting line made a decision.

"This has gone far enough," declared the starter, "Wags" Wagner.

"We agree," said the other officials.

So the first major race at the Indianapolis Motor Speedway was flagged to a stop, and the fortunes of Fisher and his associates took another plunge. The declared winner of the 300-mile Wheeler–Schebler Trophy Race was Lee Lynch in a Jack-

Fisher used millions of bricks to turn the Speedway into The Brickyard, shown here during the pace lap for the 1936 race. Rex Mays is next to the Packard pace car at right. Next to him is the great Ted Horn, and next to Horn on the outside is Chet Miller. All died racing. (Indianapolis Motor Speedway)

son, followed by DePalma, Stillman, and Harroun. Shocked fans filed slowly out of the grandstands, and a few, then as now, stood around the wrecked racers of Merz and several other injured drivers.

The Speedway did not die, though. Fisher and his group fought back once again from the bad publicity of the first major event. The track surface had been the problem. With not a thought to the great expense, they completely paved the great track with 3,200,000 bricks. Engineers had said that bricks would provide the surface needed for racing, and they were right. That's how the Indianapolis Speedway acquired the nickname The Brickyard that it kept for many, many years after.

The bricks succeeded where the crushed stone had failed. From that time on, year after year until now, attendance at the Speedway continued to increase until the 500-mile race became the largest one-day sporting event of all. The race has continued to be very dangerous, but drivers are facing the ordinary dangers of their chosen job. Spectators are protected by restricted areas and strong fences.

The greatest names in racing flew over the bricks before they were finally repaved to become the macadam surface the track now has. There were Gaston Chevrolet, Eddie Rickenbacker, Tommy Milton, Jimmy Murphy, Pete DePaolo, and Frankie Lockhart.

In those early years, the Indianapolis Speedway gained a deserved reputation as an important proving ground for brand-new automotive innovations. First tested at Indy were high-compression engines, four-wheel brakes, experimental fuels and lubricants, front-wheel drive and four-wheel drive systems, low-pressure tires, superchargers, and hydraulic shock absorbers. Racers continue to test and improve spark plugs, piston rings, turbochargers, suspension systems, and automotive design features at the Speedway. All of these in-

The vast Speedway today, shot during time trials before the 1976 race. The lower left corner is the number one turn. Between turns one and two is the fine new Speedway Museum building, housing many cars from the past. (Indianapolis Motor Speedway)

crease the safety, performance, and comfort of modern passenger cars.

As the years rolled by, new daredevils replaced the first ones. Many died on the track at Indy; others made a fortune there. Some did both.

The great Ralph Hepburn raced there. Lou Meyer, Mauri Rose, Wilbur Shaw, and Bill Vukovich—called the Mad Russian—all won the big race more than once.

As the track continued to grow in popularity, new grandstands were built until finally the entire front straightaway was enclosed by spectator stands. A new control tower was built.

Fisher, the man who had known all along that such a speedway could succeed, finally sold it to Eddie Rickenbacker. But the track languished during the Second World War. Then it became more popular than ever after the war. Rickenbacker sold it to Indiana sportsman Anton Hulman, Jr., who guided the track to the peak: the victory most wanted by every racing driver and the race best attended by auto racing fans.

Johnnie Parsons . . . Rodger Ward . . . Parnelli Jones . . . and even foreign drivers like Jim Clark and Graham Hill felt the thrill of winning the famous Indy 500. Brothers Bobby and Al Unser, and Mario Andretti, Mark Donohue, and Johnny Rutherford are modern drivers who reached the

Under the direction of the late Anton Hulman, the Indianapolis Speedway Race became the greatest one-day attraction in the world of sports. This is the exciting start of the 1975 race, with Gordon Johncock streaking into the lead just ahead of Pancho Carter (second row left) *and the all-time great A. J. Foyt* (second row right). (Indianapolis Motor Speedway)

heights at the Indianapolis Motor Speedway.

Perhaps the greatest driver of all, the legendary A. J. Foyt, thundered along the same path taken by the bumping and bouncing Marmon Wasp of Ray Harroun. Modern drivers streak past the sites of many triumphs and tragedies, adding their own triumphs and tragedies. They go more than twice as fast in racers that are scant inches off the ground and streamlined to near perfection. They earn many times more money than their earlier counterparts.

But they are all of the same breed. They are a breed of men like Carl G. Fisher and his associates, men who are told that a project is foolish or impossible, or that it is too risky, or that they should go slower and more carefully. That wasn't the way Fisher turned an old abandoned farm into what became known as "the greatest spectacle in racing," nor is it the way to drive and win the famous Indianapolis Speedway Race.

THE HOLLAND TUNNEL

Tunnel digging was hard, grueling work. Men with shovels dug the Holland Tunnel over a period of seven years. This is a cross passage between the north and south tubes, 100 feet below the surface of the Hudson River. (The Port of New York Authority)

The mile-wide Hudson River flowing between the states of New York and New Jersey was tranquil that warm, pleasant April afternoon in 1924. On the river a small pleasure boat churned its way against the gentle current. Nearby was a loaded cement barge with fifteen workers aboard.

But less than 100 feet straight down, 22 feet under the bed of the river, a terrifying drama was taking place.

There, staring upward, stood foreman David Brown and thirty-five sandhogs. Fear etched the muddy lines on each man's face. From the ceiling of the southern tube of the two mighty tunnels being bored under the river, there was a *trickle of water*. The water was coming from just behind the huge shield marking the end of that day's digging.

Quickly Brown ordered his men to attempt to shore up the area of the leak. The lower part of each tunnel was being bored through solid bedrock, but the roof section of each was really just river-bottom silt. Still, engineers had been certain that except for occasional weak spots, the twenty-two feet of hard-packed silt would hold until permanent roofs could be built. Meanwhile, silt was actually safer than rock, since silt would give warning leaks, while rock could break away and allow a sudden great flood. On this day a weak spot in the silt had been hit.

In spite of the frantic efforts of the sandhogs, the trickle of water increased to a solid stream. Before their eyes, the mud began to erode away in big chunks. Desperately they attempted to wedge shoring in place to stem the flow.

But by then there was a solid waterfall gushing into the dark, dank tunnel from the river above.

"Run for your *lives,* men!" shouted Brown. The thirty-six men scrambled back up the tunnel with a wall of water hurtling behind them. The opening in the roof of the tunnel was by then twenty feet long and a yard wide. On the surface above, the river was still calm, the scene still peaceful. But in the river's bed, there was a huge, sinking hole!

The sandhogs ran as fast as they could. One hundred feet . . . two hundred . . . the water boiled up the gradual slope of the tunnel behind them. Finally, dashing toward the Canal Street end of the tunnel, they moved high enough that the slope and the air pressure, always pumped into a tunneling operation, slowed and finally stopped the flow of water. But most of the tunnel was flooded.

At that moment the tranquil scene on the river's surface was suddenly shattered. Without warning a great geyser of water shot high into the air from directly under the sterns of the pleasure boat and the barge. The boat bounced up and down, rocking violently from side to side. The barge was lifted by its stern, dumping and sliding workers and cargo into the water.

Air pressure in the tunnel pressing against the

water had forced it suddenly to back up. As though from a blown-out tire, the water had shot upward through the river and into the air.

Fortunately, nobody was injured that day. Sandhogs were able to save the project by dumping barge-loads of clay onto the exact spot of the blowout from above, while increasing the pressure of air in the tunnel to force out the water. Work resumed in the tunnel three days later.

Skeptics had insisted from the beginning that the vehicular tunnel under the Hudson was a fool's dream, that it would prove to be impossible to build. A short tunnel was one thing, but this was to be the world's longest tunnel for cars. Still, traffic between lower Manhattan in New York and Jersey City, New Jersey, had increased so much by the 1920s that either a bridge or a tunnel was deemed absolutely necessary. Officials in both states decided a tunnel would be best.

But who could take on the task of digging such a tunnel? Who had the skill? It would be nearly two miles in length, with more than a mile directly under the great Hudson River. For safety the roadway would have to be nearly 100 feet below the surface of the river at high tide. More than one-half million cubic yards of earth, rock, and silt would have to be excavated, every foot of it in dark danger and every minute with the threat of bends and blowouts and drowning. Buildings and entire neighborhoods would have to be removed to provide for entrance and exit plazas, and two huge ten-story buildings would have to be built to house fans and ventilation equipment to keep the air in the tunnels from poisoning motorists.

Although older engineers of national and international reputation in tunnel construction were on the list considered by the officials of the New York State Bridge and Tunnel Commission and the New Jersey Interstate Bridge and Tunnel Commission, thirty-six-year-old Clifford M. Holland, a brilliant young civil engineer, was chosen to construct the tunnel.

Young Holland was no stranger to tunnel digging. Slimly built, tanned, and with the air of a man who knew exactly what he was doing at all times, Holland had sold the commissioners on his plan even though they had already tentatively selected Panama Canal builder General George W. Goethals to build the tunnel.

The famous Holland Tunnel in New York might just as easily have become the Goethals Tunnel but for the persistence of Cliff Holland, a graduate of Harvard who went to work as a tunnel engineer for the Public Service Commission of New York. In fourteen years Holland had put four tunnels under New York's East River. When Holland talked of tunnels, the officials listened carefully.

Both Goethals and Holland suggested the standard construction method with shielding and compressed air. This is a tunnel-boring process in which sandhogs use huge braces and shields to prevent cave-ins and in which compressed air creates enough inside pressure to prevent leakage. But Goethals submitted a plan that called for forty-two-foot diameter tunnels. Holland's plan required twenty-nine-foot diameter tubes, and he spoke of his own conservatism in the dangerous art of tunnel digging.

"As the diameter of a tunnel increases, the difficulties increase very much more rapidly," he cautioned the commissioners. His youthful face and appearance were against him, but he continued with all the sincerity he could muster. "With my appreciation of these difficulties and in the present state of the art of tunneling, I don't advise tackling any tunnel bigger than twenty-nine feet. When we have built one that size, it will be time enough to build one bigger."

Holland looked around at the different members of the commissions. He wanted the tunnel job with all his heart. "Besides, gentlemen, I don't see the advantage which the larger tunnel has over the twenty-nine-foot size, although there are many claims for it." He pointed out that Goethals planned to use concrete alone for the lining of his tunnel. "It won't hold up in the conditions found under the river," Holland insisted. He planned to use concrete and cast iron, a combination he called "wonderfully well-suited for the purpose."

So Clifford M. Holland, a father of four daughters, was awarded the job of chief engineer on the largest tunnel in the world. The commissioners were convinced that he would work the hardest and do the best job.

Years later, on October 8, 1924, a small item appeared in *The New York Times* under the headline TUNNEL ENGINEER GETS A VACATION. The short article noted that Holland had been awarded a one-month vacation from his job as tunnel engineer due to his "continuous devotion, night and

Clifford M. Holland is second from the right at this New York groundbreaking ceremony for the tunnel on October 12, 1920. (The Port of New York Authority)

day, during the past five years." Nobody knew when Holland took the job that he would literally work himself to death on the tunnel. He died at age forty-one, only two days before the two ends of the first tunnel met deep under the Hudson River.

Engineer Milton H. Freeman took over from Holland, but he too worked day and night and finally collapsed and died. The Manhattan plaza of the tunnel is officially named Freeman Square, a fact sometimes overlooked by city mapmakers.

Finally, engineer Ole Singstad finished the long, difficult tunnel.

But it was Holland who recognized from the start that although a tunnel could be dug by enough men, another problem was less visible and more important. What about air? Little was known about the hazards of breathing exhaust fumes from automobiles.

Other tunnels such as the Blackwall and the Rotherhithe, two famous but shorter ones under the Thames River in London, were ventilated by a natural flow of air. The great tunnel at Pittsburgh, Pennsylvania, was ventilated by a center tower that forced fresh air down into the middle of the tube. But the Hudson River was one of the greatest waterways of commerce in the country. Huge cargo ships, coal carriers, and ferryboats steamed up and down and crisscrossed the river constantly.

A full-scale model of the proposed vehicular tunnel under the Hudson River. There were to be two such tubes: the traffic in one going east, the other west. Note the flow of fresh air coming in through the curbs and exiting through the ceiling to carry gases out. (The Port of New York Authority)

Any ventilation tower built in the middle of the river over the center of the tubes would be a hazard to navigation. Yet Holland knew that artificial forced-air ventilation would be essential in the long tubes he planned to dig.

So Cliff Holland set in motion a series of unique tests to determine whether or not people could live while driving autos through long tunnels, even *with* ventilation. He hoped to refute the many internationally respected engineers, contractors, and scientists who insisted that the tunnel itself was possible, but that it would be a deathtrap for anybody who tried to use it after completion.

Holland personally supervised the experiments, beginning the twenty-four-hour workdays that would eventually claim his life. First he obtained an abandoned coal mine near Pittsburgh, where tracks were constructed so that cars could be tested in confinement. Each entrance to the mine was sealed with a bulkhead so that precise air readings could be charted. Holland tested empty autos, full autos, and trucks, all traveling at various speeds. When he was finished, he had an accurate picture of the amount of gases that would be freed in the tunnel under the Hudson under any condition.

With these figures he was able to show when the air would become dangerous and how quickly it had to be changed for health reasons. At the same time tests were being conducted in the chemical laboratories at Yale University on how polluted air affects first guinea pigs and then people. Today, testing polluted air is common, but in Holland's day these were the first studies of what smog and other impurities can do to a human. Holland finally determined that 4 parts of gas (auto exhaust) to 10,000 parts of fresh air would be dangerous. Thus, he knew exactly how much fresh air would be needed to keep his tunnels safe.

Finally, he built a miniature tunnel at the University of Illinois to the exact scale of the tunnel he planned to dig under the Hudson River. With mathematical accuracy he proved his calculations. He concluded that four stations, two on each side of the river, would be needed for ventilation. He planned two ventilation stations onshore on each side, and two stations at the end of piers 1,000 feet out into the river on each side. The latter two would pump fresh air down to the tunnel, then out each of the land entrances. The other two would pump air to the center of the tunnel.

The air is changed in the Holland Tunnel once every forty seconds, or three times every two minutes. It is supplied by immense fans in the four buildings. The fresh air travels under the roadway and up through vents in the curbing. It then travels up through vents in the ceiling and away, carrying gases with it. It is said that the air in the Holland Tunnel is cleaner than the air in midtown New York City.

Gradually, inch by inch for two long years, the shields were moved slowly under the river. On some days only an inch or two was gained. Eating through rock and mud, the men extended the length of the tunnels, two crews moving toward each other.

Holland faced and solved problems day and night. He was always available to his workers, and often they would call him from bed to handle a situation on the job. He made constant decisions on ventilation, cave-ins, blasting difficulties, shield troubles, labor disputes, compressed-air emergencies, the use of materials, earth-type tunneling changes and even flooding in the tubes.

This is the south tube shield at the face of the digging. Slowly this shield was pushed deeper and deeper under the river. (The Port of New York Authority)

One by one, thirteen workmen lost their lives as the tunneling progressed. Every day, every worker among the hundreds had to endure decompression treatments to prevent the bends that could result from spending too much time under more air pressure than is normal. The health of the chief engineer also began to fail. He was warned by doctors that his heart could not take the strain of his long days and nights without sleep or rest. Still, he worked on. Co-workers could come to him at any time with a problem, and from his vast store of knowledge on tunneling he would solve it.

"I am going into tunnel work," he had said upon

graduation from college years before, "and I am going to put a lot more into it than I'll ever be paid for." Holland was right. Though he was offered jobs paying far more than his tunnel work, the great vehicular tunnel under the Hudson River became an obsession with him.

Finally he was forced to take a vacation. Tired, wan, and weary, he went to Battle Creek, Michigan, to regain his health. Work on the tunnel went on. Three weeks later a friend visited him and came back to New York to report that Holland's health appeared much better, that he was anxious to return to the tunnel. For the "hole through" was due, the time when the crews in the tunnels would meet deep under the riverbed. Great ceremonies were planned for the occasion, and Holland wanted to be present.

But as sandhogs were preparing the final blast deep under the river, Clifford Holland died. He did not live to see his greatest dream come true.

Sandhog electrician A. F. Templin stood back as the final blast was fired; then he squeezed his body through in the damp, tomblike darkness far under the river to grasp the hand of a worker from the other side. It had been months and years of toil, of chipping away at bedrock and fighting the oozing silt and mud, but the tunnel had been opened.

The celebrations were canceled because of Holland's death, and work went on. The hole through became just another part of a routine day's work, but the tunnel that had been called impossible only a few years before was a reality. Men were passing through regularly. With both tubes through, the work of finishing and tiling proceeded while the construction of entrance plazas, toll booths, and ventilation systems began.

Nor did testing in the tunnel ever stop. Critics found a new problem to shout about. "If a car catches fire in the tunnel," they insisted in news-

The air is changed inside the Holland Tunnel every forty seconds by huge fans. The one shown here is in the New York station. (The Port of New York Authority)

papers, "everybody in there will be burned to death. The fire will leap from car to car, and nothing can stop it."

It was another example of "tunnel fear" that critics attempted to spread. Terrible accidents, death from foul air, the bends, entrapment, and other fearful things were predicted throughout the building of the tunnel, but one by one engineers allayed the fears.

An old touring car was put in the middle of the tunnel during the fire scare. Six gallons of gasoline were poured over it and allowed to run down the roadbed. Then the gasoline was ignited. Fire leaped up the tiled walls and licked at the concrete ceiling.

Only at that instant did the tunnel fire force, by means of a fire alarm, know that the experiment was taking place. In only three and one-half minutes, using a fire lane and the best equipment of the day, the firemen extinguished the blaze. Other nearby vehicles were undamaged.

Upon leaving the tunnel, witnesses noticed another old, burned-out car in a field near the entrance. The firemen explained that they had done the same test earlier that very morning . . . just to be certain.

Years later, on May 13, 1949, the worst possible thing happened in the tunnel. There had been many quickly controlled fires in vehicles, but on this day a truck loaded with chemicals crashed, caught fire, and exploded in mid-tunnel. Twenty-six persons were hurt, one fatally, and $500,000 damage was done. But the tunnel fire department still prevented a major tragedy, for the tunnel had been crowded with cars.

Now, dangerous cargoes are prohibited inside the tunnel.

On November 12, 1927, thousands of people walked through the tunnels, from one end to the other, on the day before the dedication.

On November 13, 1927, the world's largest vehicular tunnel, the only one with air conditioning, opened with a charging, horn-blowing stream of thousands and thousands of "first timers" roaring their cars through. The trip from Manhattan to Jersey City had been cut from thirty minutes to only three. The north tube was westbound, one way. The south tube was eastbound, one way. Eighty-four huge fans cleaned the air as more and more cars continued to stream through.

At the dedication ceremonies, a long moment of silence gave tribute to Engineers Holland and Freeman and the thirteen men who had died on the job. The tunnel was then officially named Holland Tunnel.

Cars have passed through the great tunnel at rates of up to 2,400 per hour ever since. What was once a thrilling journey of a lifetime has become commonplace as the tunnel has taken its place in the transportation systems of New York and New Jersey.

The tunnel has been a trap for many criminals who race in one end and are caught by tunnel police before they can exit out the other. Once a tank truck full of sticky syrup developed a leak, and before it could be stopped, most of the tube was an inch deep in molasses. Traffic was halted for an hour while the mass was sanded down.

The interior of the Holland Tunnel, with a tunnel policeman on the narrow ramp alongside the roadway. The vehicle in which the policeman is standing is one specially made for traffic control in the tunnel. It runs along the narrow catwalk. (The Port Authority of N.Y. & N.J.)

There is an average of twenty-five mechanical breakdowns every day in the tunnel, but quick handling by police trained in auto mechanics and fast action by the tunnel tow service almost always prevent major traffic jams. Most breakdowns are removed from the tunnel in less than five minutes. Tunnel maintenance is accomplished with specially designed equipment such as snow melters, lane marker line painters, tile washers, and tunnel police vehicles.

Motorists pay a toll to drive through the tunnel. More than twenty million cars have used the tunnel in one year. What cost a few million dollars to build has earned back its investment several times over, and continues to earn, helping to pay for other transportation projects.

The name Clifford M. Holland, one that might have been forgotten even though the man accomplished seemingly impossible tasks with a drive and dedication unmatched by most people, has taken its place with those of the great project engineers of the world. For as long as the Holland Tunnel is there, Clifford M. Holland will be remembered.

And even longer, for the techniques he developed and used to construct his great tunnel in New York are still used in tunnels throughout the world.

MOUNT RUSH ORE NATIONAL MONU ENT

The Needles of Mount Rushmore in South Dakota before the sculptor began work. He took initial measurements by running a single line up to a tower at the top. (South Dakota Tourism)

There is a beautiful likeness of the head of Abraham Lincoln in the rotunda of the Capitol at Washington, D.C. The head was carved in loving detail from a six-ton block of marble by a superb sculptor. There is another statue of a seated Lincoln in front of the county courthouse in Newark, New Jersey. So striking is the statue of the sixteenth president that it appears he could almost stand and stroll away.

In the Metropolitan Museum of Art in New York City is *The Mares of Diomedes*. There is a series of figures of the apostles of Christ in the Cathedral of Saint John the Divine in New York City. There are statues of Philip Sheridan and James Smithson in Washington, D.C. There is a memorial for the Gettysburg battlefield in Pennsylvania. All of these were done by the same sculptor, Gutzon Borglum.

Thousands of miles from these beautiful works stand the spires of one of nature's great sculptures, the towering Needles in the Black Hills of South

Dakota. A part of Mount Rushmore, these granite pinnacles are as ruggedly beautiful and inspiring as any monument built by man.

Some people, in fact, felt they were an ideal place to carve great figures of outstanding westerners like notable Sioux Indians who once lived in the area or western heroes like Lewis and Clark, John C. Frémont, Jedediah Smith, Jim Bridger, or Buffalo Bill Cody. There, far up in the sky, they could gaze down on the magnificent country they had helped to settle.

But other people, those with affection for the area, said, "Man makes statues, but God made the Needles. Let them alone!"

Finally, in August 1925, the sculptor responsible for the infinitely fine statues in the East rode on horseback into the wild Black Hills country, near the exact geographic center of the United States. He came upon Mount Rushmore. He stared up at the towering rock. If ever an artist had a chance to leave a memorial that would last for countless thousands of years, this would be the raw material.

Borglum, who at that time had been working on the huge Stone Mountain carvings in Georgia, decided to stop work there and concentrate on this one last great carving. But he felt that the monument at Mount Rushmore should be to great American figures rather than to just great western figures, men national in character rather than regional.

Finally chosen were George Washington, Thomas Jefferson, Abraham Lincoln, and Theodore Roosevelt. Permission for the amazing project was granted by the federal government, which owned the land, and preliminary work began. This brought the first of many, many problems for Borglum, who had hired his son, Lincoln, to work with him on the project.

The faces for "America's Shrine of Democracy" were to be sixty feet tall, twice as tall as the Great

Hand-operated winches lowered workmen over the side of the figures. Note the huge crack in the side of Jefferson's nose, a rock flaw which forced Borglum to move *the nose, before it was finished.* (South Dakota Tourism)

Sphinx in Egypt. Borglum's first studio model of Mount Rushmore had Jefferson at the left, followed to the right by Washington, Lincoln, and finally, Roosevelt. That is not how the monument was carved, although Borglum felt at first that this was a fitting and chronological order for the great faces. Because it would dominate the group, the Washington face was carved first and dedicated on July 4, 1930.

". . . The great face seemed to *belong* to the mountain," said Borglum. "It took on the elemental courage of the mountain around."

Following his original plan, Borglum then blocked out the face of Jefferson on the rock to the left of Washington. That rock is still there, partially flattened but mostly as nature left it, for flaws in the structure and a thinness of rock surface forced a change in the basic design of the monument.

Jefferson was then positioned to the right of Washington, and the features were blasted into the rock in 1934. But once again a problem arose. When a flaw develops while he is carving on a smaller block of granite, a sculptor can select another, unflawed stone. But what do you do with a *mountain*? A great crack had been found through Jefferson's nose.

Moving the 6,000-foot-high head appeared to be the only answer. Fortunately, Jefferson's nose could be moved, and this shifted the position of the entire head enough. Today, the flaw in the rock, almost invisible, runs just to the side of the great nose. In the generations to come, Jefferson's nose will *not* fall off.

To the far right was a huge flat surface on which was to be carved an historic tablet. Borglum envisioned that the words on the tablet, when gilded, would be visible three miles away. Some years before, when he was President, Calvin Coolidge had agreed to write the words about what Borglum considered the most important events in the history of the United States.

These were, according to Borglum, the Declaration of Independence, the adoption of the Constitution, the Louisiana Purchase, statehood for Texas, settlement of the Oregon Territory dispute, statehood for California, the end of the Civil War, and the building of the Panama Canal.

Whether or not President Coolidge agreed, the manuscript he submitted was far too long and too detailed, in Borglum's opinion, to carve on the side of a mountain. It would be too complicated and would overshadow the faces. Instead of asking Coolidge to shorten the manuscript, Borglum merely edited it himself.

Coolidge, like any other fledgling writer, took offense at this and refused permission to use the shortened version. He also refused to write a new one.

So finally, Borglum merely blocked in the head of Lincoln in the space reserved for the tablet and proceeded to blast away. A huge "1776" already carved as part of the tablet was unceremoniously blown away for Lincoln's forehead.

All of this brought great publicity to the project, and thousands of people visited to watch the work in progress.

Oddly, as the monument progressed, the skilled Borglum seemed to be able to "move" the figures

To the far right was a flat surface for a tablet, but Lincoln was finally carved there. Many feel this is the finest carving of all. A vein of silver was found in Lincoln's nose. (South Dakota Tourism)

to make them more artistically perfect. At one point during the carving of the Lincoln figure, he decided to "move" Washington for a more pleasing appearance. The new position was accomplished by merely shifting Washington's left shoulder back fourteen feet. The shift gave the monument an entirely new appearance, that seen today.

The Lincoln head was dedicated on September 17, 1937, and the squeezed-in Roosevelt head on July 2, 1939. Until Borglum's untimely death following a heart attack in March 1941, he was doing the final blending and finishing work on the figures. So he did see the great project of his dreams before he died.

How does a sculptor carve a high, rugged mountain face?

It is simply a matter of removing unwanted rock. The faces, according to Borglum, were there all the time. If the correct pieces of rock are removed, what is left will be the monument as the sculptor has designed it. More than 400,000 tons of rock were removed from the Needles of Rushmore, much of it still in a jumble at the base of the cliff.

At first a perfect scale model of the final monument was built in the sculptor's studio. It was correct in every detail and took into consideration the rock face upon which the work would be done. Meanwhile, the rock surface of the mountain was blasted away with dynamite until the deep fissures and cracks were eliminated and solid granite was reached. That was the sculptor's working surface.

During this process, the Rushmore model was changed nine times in the studio. Before a suitable rock surface was finally found, Washington's chin was moved back thirty feet and the entire Jefferson head was moved back sixty feet from where it was originally planned to be. Each time the model was changed, unity in composition had to be restudied.

Pointing, one of the first and most important operations, indicated to workers exactly where to blast and how much rock was to be removed. Borglum's son, Lincoln, was in charge of this process. At the top of each head on the model in the artist's studio was a small boom. On each of the large figures on the mountain there was a corresponding large boom.

A circular plate, small on the small boom and large on the large one, was attached to a line from

Pointing indicated to workers exactly where and how much rock was to be removed. A red mineral called allanite was discovered in Roosevelt's cheek, but it didn't stop the carving. Here a workman is examining the mineral deposit. (South Dakota Tourism)

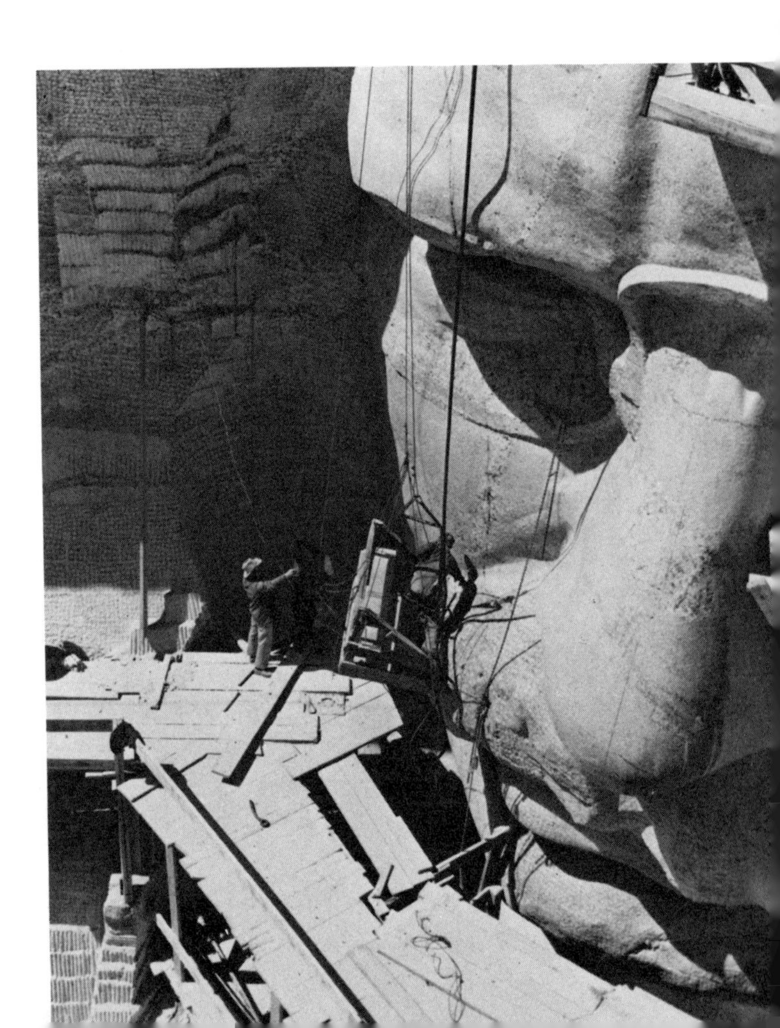

each boom. Degrees were marked on each plate. Plumb bobs could be set out ten inches on the small boom and ten feet on the large one. If the tip of the bob on the small boom was lowered thirty inches to touch the point of the nose on the model, then the tip of the bob on the large boom was lowered exactly thirty feet to indicate to workers the tip of the nose on the monument.

In this manner the faces could be matched exactly to the model, only much, much larger.

Rock for the actual figures was removed by careful blasting. Precision was required to remove the right amount of stone from the right place at the right time without going too far and without damaging nearby stone already carved. Borglum's employees became so skilled at this work that, by blasting, they could block out a nose to within *one inch* of the final shape and surface. They could shape lips, grade the contours of a neck and forehead, and cut round areas, such as eyeballs. Then holes were made with rock drills to the exact shape and surface of the final figure. A bumping process with a rounded tool and a jackhammer pounded away the drilled surfaces down to the surface desired. Most of this work and all of the blasting work were done from high scaffolding and hanging workers' cages that took the breath away from visitors below.

Eyelids and pupils were shaped by drills under Borglum's constant supervision. Borglum himself used hand tools on the faces, to give them "life."

There is a shaft of granite, for example, in the pupil of Lincoln's eyes, left to capture the sun and give the eyes "sparkle" from a viewing distance. Visitors often remark at the lifelike appearance of the great eyes.

Every year millions of visitors stand in awe, staring upward from an observation room far below the great faces. But most do not know that the vast monument was never finished. Nor will it ever be

Workmen used rock drills to honeycomb the granite down to the surface desired. Then rock slabs were wedged off, and finally the surface was bumped smooth before it was polished. (National Park Service)

finished, at least not according to the plans of its creator.

The original plans called for the figures of the four presidents to be finished to the waist. Washington, in fact, exhibits shoulders, a coat collar and lapels, a neckpiece—and he even has a button on his coat.

Gutzon Borglum also envisioned and planned for a mighty rock hall in the *back side* of the figures. This huge hall was to contain 360 feet of wall space, with lighted 30-inch insets holding records of the accomplishments of the Western world. There would be displays about science, art, literature, inventions, medicine, and the United States government. A great gold-plated bronze, bas-relief was to stretch around the entire hall. It would show the "Adventure of Humanity."

The grand hall was to be entered through cast glass doors more than 22 feet in width. Over the doors would hover a huge 38-foot eagle. On 49-

foot-tall pylons on each side of the entrance there were to be 30-foot-tall flames.

Cut into the stone at the entrance were to be the words *America's Onward March,* and beneath that the words *The Great Hall of Records*. Borglum was sure that the hall would be safe and that the records of America could best be preserved there.

Leading up to the hall was to be a long flight of steps carved into the rocks. Invisible from the front of the monument, the stairs were to be 15 feet wide and 800 feet long.

They became the stumbling block of the remainder of the project, possibly the final straw that ended the dream. Advisers suggested that the stairs be the first part of the back side of the project, and that they be carved by the Civilian Conservation Corps under Borglum's direct supervision. The CCC was an army of volunteer workers paid by the government during the Depression. Many parks and other such projects were built by the CCC, and many men happily joined the CCC and did this type of work during the days when American families were starving from lack of income.

But Borglum, an artist to the end, refused flatly. He insisted that highly skilled workmen, even sculptors, were needed for the carving of the stairs.

"It must be designed with taste and monumental dignity," he said, so he designed it all himself. "It must be built in harmony with the big character of the work," he then said, refusing the offer of CCC workers. The stairs were never built.

But a tunnel, the beginnings of The Great Hall of Records, was started. It is still there in the back of the mountain behind the stone faces.

Lincoln Borglum took over after his father's death. He worked on Roosevelt's face, Lincoln's head, Jefferson's collar, and the collar and lapels on Washington's coat. He agreed that the front side of the monument should then be considered complete.

"I do not think any more should be done on the figures," he wrote. He felt they were just as effective as they would be if each had been carved to the waist.

Lincoln Borglum was probably saving his persuasive powers for the back side of the monument. He felt very strongly that the hall and the stairway should be completed. He fought for this, feeling that it was necessary to preserve the intent of the stone faces.

He made the point that without the hall, "the memorial will become a puzzle to the people a few thousand years from now."

He was probably right, although money had run out by then anyhow. Even today we are no longer certain of the specific intents of the builders of the Great Pyramids or exactly how they were built. We certainly have no understanding of many great statues and markings in the earth from our dim past. It is almost certain that someday in the distant future the great faces on the mountain will puzzle visitors. They will wonder who the men were and they will marvel that a civilization as primitive as ours found the ability to carve them.

For the great Mount Rushmore Memorial will last into the dim future. High granite of the type in the Needles erodes less than *1 inch* every *100,000 years*.

The great five-story faces in the rock will be there virtually forever.

Mount Rushmore National Monument as it appears today. In the distant future people may wonder whose faces they are and who ever had the skill to carve them. (National Park Service)

THE EMPIRE STATE BUILDING

No construction man had ever worked that high before. It would be a building project of stomach-wrenching heights above the streets of New York City. Sky boys would step from thin girder to girder 1,000 feet above the concrete and rise from one level to another at the end of a cable. They would eat their lunches with feet dangling over an abyss that would turn the nerves of most people to jelly. They would handle bucking rivet guns and hot welding torches at dizzying distances above the ground.

And they did. As the mighty Empire State Building reached into the clouds, workmen walked casually about on narrow beams with smiles on their faces, proud of what they were accomplishing and with not a thought of the frightening chasm below.

In May 1931, the *Literary Digest* said of the workers on the Empire State Building, "Like little spiders they toil, spinning a fabric of steel against the sky. Crawling, climbing, swinging, swooping . . . weaving a web that was to stretch farther heavenward than the ancient tower of Babel, or than all of the elder towers of the modern Babel, as Manhattan has sometimes been tagged."

With flawless lines and striking beauty, the great building, the tallest in the world when it was built, frequently does project higher than the clouds. Visitors on the observation deck on the 102nd floor often see rain falling from clouds below them. And something even more exciting. The building became the greatest lightning rod ever built. It takes about 1/6000 ampere to kill a human. The Empire State Building was once hit

Photographer Lewis Hine took this dramatic photo of two nonchalant sky boys resting in the girders of the tower of the Empire State Building, more than 1,000 feet above the sidewalk. Hine, of course, was even higher on the girders. (Avery Architectural Library)

with 200,000 amperes nine times in a span of twenty minutes during a thunderstorm. There was no damage to the building or people, of course, since the vast steel framework merely absorbed the electricity.

The maximum height of buildings was once limited to a few stories. Builders used bearing-wall methods of construction, in which the walls supported the structure. But with the skeleton-frame type of construction, buildings began to shoot upward. In this type of building, the entire structure is supported by a steel framework of columns, girders, and beams. Inner and outer walls enclose only space.

The first skeleton frame building was the huge Home Insurance Building in Chicago, built in 1884. Construction men reached a mighty new height of *ten stories* on that building. And after that job they knew the sky was the only limit. At least on paper.

New Yorker Alfred E. Smith was president of a corporation formed to build the tallest building in the world. Smith was just the man for the job. An American political leader, he had been governor of the state of New York four times. He was once candidate for president of the United States, although he lost the election to Herbert Hoover. Nicknamed The Happy Warrior by President Franklin D. Roosevelt, Smith was a man who loved mighty projects and who usually got what he wanted one way or another.

He had been born in a slum in the East Side of New York and became a newsboy and a fishmonger after leaving school at the age of twelve. At age twenty-two he entered politics, and from then on it was up for Smith, all the way to prime mover of the Empire State Building.

The Empire State Corporation finally approved the *sixteenth* plan submitted by architect William Lamb, and everything was ready.

Meanwhile, New Yorkers watched with mixed emotions as the grand old Waldorf-Astoria Hotel was being demolished to clear the corner of Thirty-fourth Street and Fifth Avenue. The hotel, built by William Waldorf Astor and his aunt, was one of the last of the true luxury hotels. Parties to stagger the imagination were held in suites at the Waldorf-Astoria. Entire floors were redecorated to please certain rich guests. But in spite of its worldwide reputation, not many people asked for souvenirs from the hotel, and most of it was finally carted away in trucks, loaded onto barges, and dumped into the ocean five miles off Sandy Hook, New Jersey, where it still lies.

Within a month after approval of the architect's plans, huge shovels moved in to begin digging a massive hole at the famous corner. It was February 1930, and the country was in the grip of a great depression. Still, financing had been arranged, and truckload after truckload of rock and dirt rumbled away down Fifth Avenue. Even today visitors ask if the tremendous weight of the great building might not eventually cause it to sink into the bedrock upon which it is built. But the fact is that the total weight of the dirt and rock hauled away approaches the weight of the building itself. The bedrock isn't working much harder to hold up weight than it ever did.

Little did workingmen or designers know then that many years later, long after the building had proved its solid strength and durability, it would face one of its greatest tests. On Saturday, July 28, 1945, an overcast morning with a light drizzle falling, fifteen years after the block-long hole had been started, an army B-25 bomber speeding at 200 miles per hour crashed directly into the side of the building between the seventy-eighth and seventy-ninth floors on the Thirty-fourth Street side.

When it hit, the great building rocked and shook, and people inside believed that an earth-

The seventy-ninth floor after an airplane crashed into the Empire State Building in 1945 (New York Times)

quake had struck. A huge eighteen-by-twenty-foot hole was ripped into the gleaming outer wall of the building, and flames roared up the side from smashed fuel tanks.

One of the massive engines from the airplane skidded down a long hall, crashed out through the other side of the Empire State Building, and fell down, down to smash through the roof of a twelve-story building next door. The resulting fire destroyed a well-known sculptor's studio in that building's penthouse.

Far above, flaming gasoline was flowing down stairwells all the way to the seventy-fifth floor, causing major fire damage. Parts of the airplane flew up as high as the eighty-sixth floor observation deck. The massive steel girders that had been riveted into place so many years before by the sky boys at the seventy-ninth floor were bent inward eighteen inches.

The other engine of the twin-engine bomber

skidded into an elevator shaft and fell all the way to the subbasement, 1,000 feet below, landing on top of an elevator car. In doing so, it cut the cables, and another elevator car fell to the basement as well. This car was carrying two women.

Far down below ground level, in the lowest basement of the building, firemen cut through a wall and through the side of the second crushed car. They fully expected to find the occupants dead after their terrifying fall. A young coastguardsman, a hospital apprentice selected because of his small size, was chosen to crawl through the hole to examine the bodies.

"Thank heaven," said one of the women, "the navy's here." Miraculously, though burned and seriously injured from the fall, both women survived.

Others were not so lucky.

Fourteen people in the building died, including the three-man crew of the airplane. Twenty-six more were injured. Fire ravaged two complete floors of the building and damaged several other floors. Other buildings were damaged. It took one million dollars to repair everything.

But the structure of the Empire State Building was sound, and it survived the terrible day without permanent damage. It was repaired, and today all evidence of the crash is gone.

The great hole on Thirty-fourth and Fifth was finally excavated, and on March 6, 1930, the area was ready for the steelmen. Deep footings called pile foundations had been set down into the bedrock of Manhattan Island. On April 7 the steelwork began, and New Yorkers watched in amazement. Like a child's erector set on a grand scale, the boxlike structure called The Avenue of Girders climbed.

Five workers fell and died before the steel reached the lofty 102nd-story level; fourteen died in all, but work continued. Workmen moved about as though they were at ground level, and New Yorkers watched in awe as the tallest building in the world moved ever upward at a surprising pace.

Steel was manufactured and cut to size at the steel mills in Pittsburgh. With military precision it was hauled first to a supply depot across the bay at Bayonne, New Jersey, and then by barge to Manhattan Island. Trucks picked up the steel beams and carried them to the construction site. Each beam had been marked to show exactly where it was supposed to go in the building.

Nine derricks powered by electric hoists lifted the girders into position, where they were first bolted and then riveted permanently into place. In his fine book *The Empire State Building* (Harper & Row), Theodore James, Jr., explained the precision that made possible the speed of construction:

> The construction went forward with absolute precision. On each floor, as the steel frame climbed higher, a miniature railroad was built, with switches and cars, to carry supplies. A perfect timetable was published each morning. At every minute of the day, the builders knew what material was going up on each of the elevators, to which height it would rise, and which gang of workers would use it. On each floor the operators of every one of the small trains of cars knew what was coming up and where it would be needed. Down below, in the streets, the drivers of the motor trucks worked on similar schedules. They knew, each hour of every day, whether they were to bring steel beams or bricks, window frames or blocks of stone. The moment of their departure from the storage place, the length of time allowed for moving through traffic, and the precise moment of arrival at the site of the building, all were calculated, scheduled, and fulfilled with absolute precision. Trucks did not wait, derricks and elevators did not swing idle, men did not wait. So perfect was the planning that workmen scarcely had to reach for what they needed next.

As the building frame rose, the hoist engines were lifted up outside the building to successively higher floors. As in the later Gateway Arch in Saint Louis, Missouri, the lifting devices climbed with the project.

When a beam of steel would arrive at the construction site, a fitting-up gang would get it to the correct derrick and attach it. Then a raising gang would lift it to where it was needed. Then, after the sky boys had located the beam in the frame-

During construction, sky boys bolted and riveted into place massive steel girders. (Avery Architectural Library)

work, heater gangs would prepare rivets, bucker-up gangs would prepare the riveting, sticker-in gangs would insert the rivets, and riveter gangs would pound them home. This group of people formed one team of workers.

There were thirty-eight such groups working on the great building's frame, totaling more than 1,000 workers at any given time.

As the gangs completed the framing for a floor and moved on up, cement masons and other stone-workers would move in to enclose that floor and prepare it for finishing workers. They would come in and partition, wall, and window the floor. Onward and upward the Empire State Building climbed into the New York sky, changing and marking the skyline of the vast city with the familiar outline it now has.

Nonchalantly the workers continued to defy gravity, jumping from one beam to another. At lunch as they sat with legs dangling, they laughed and talked as though they were at their dinner tables at home. Back at work, they would casually look upward at a new beam swinging into place while below waited eternity.

Some weeks the high iron men pushed the steel framework up a full four and a half stories. It took a little over a year to build the Empire State Building, an astounding feat when structures such as the Taj Mahal are considered. That beautiful building, though much smaller and far less complex than the Empire State Building, took 20,000 workers more than twenty years to complete.

As the steel skeleton rose and the floors were being enclosed, Smith and his associates were renting space in the building. Soon it would be open for business.

Nearly two million people per year visit the Empire State Building which, with its television tower, is 1,472 feet tall. The building contains 6,500 windows, 1,860 steps to the 102nd-floor observation platform, more than 350 miles of telephone cable, 70 miles of water pipes, and 7 miles of elevator shafts. The elevator cables alone would stretch from New York to Florida, and at the ends of the miles of telephone cables are a total of 18,000 telephones.

Lights atop the building can be seen from eighty miles away. During a heavy wind, the building sways slightly and actually makes creaking sounds as its mighty steel skeleton gives with the strain. But it doesn't sway as much as rumors insist. Although some people claim that the sway is up to eighteen feet and that you can barely hold your footing in the upper floors during a heavy wind, instruments have shown that the greatest sway was once an inch and a half off center. And that was

Workers casually ate their lunches with legs swinging from girders 100 floors above the city, then went back to work without a care, proud of what they were building. (Avery Architectural Library)

Building inspectors followed the workers, measuring and checking to make certain that the grand building was solid and almost everlasting. But in so doing, they took what appeared to be terrible gambles. (Avery Architectural Library)

when a rare 100-mile-per-hour wind gust hit the building square on.

In November 1930, a dirigible mooring mast was installed on top of the building tower. This mast, it was hoped, would lead to regular dirigible passenger stops at the building. Passengers were to descend from an airship in an enclosed staircase leading to a tower elevator, while the airship was to be moored by its nose to the building tower and to swing free otherwise. No wonder the plan never worked out.

New Yorkers had watched in amazement as the skeleton of the building climbed the first dozen or so floors, but then they proudly began to point at the massive structure. They had never had a doubt, they said, and they drove for miles to show out-of-town friends the progress on the building. Even today visitors can take a moment from their own enchantment and catch native New Yorkers glancing up with pride at the marvelous building. The shimmering facade of chrome-nickel-steel alloy on the building never tarnishes or dulls. Even from the far New Jersey shore, the setting sun is reflected from the finish and turns the spire a blazing red.

Though other, newer buildings are now higher in the sky, the Empire State Building represents the first in the world. Still more beautiful than any other, it marks the ability and skill of the people of New York, the courage to take a chance on a project that was proved as it climbed.

The final scaffolding was removed from the rounded tower of the building in March 1931. Inside, the water pipes were being tested to be sure there were no leaks. Rugs were being laid, walls painted, and other finishing work was being done. The building was not yet open to the public, but craftsmen swarmed through the halls, paneling, tiling, and doing other last-minute work. For the grand opening was approaching.

Across the vast city of New York, amateur astronomers had turned their telescopes from the moon to the mooring mast atop the building. It was *the* thing to watch, although no airship moored there.

In April the Empire State Building was awarded the Medal of Honor from the Architectural League as the best construction of all. A few people still doubted the need for skyscrapers, but Al Smith continued to speak of the size of the building, insisting that time would justify its massive space and height. Because of the Depression, it had become obvious that the building would not enjoy full occupancy initially, and in fact it was many years before it was filled with tenants. Meanwhile, the thousands upon thousands of visitors willing to

The magnificent Empire State Building shortly after it was opened. It is one of the most beautiful buildings ever constructed. (Museum of the City of New York)

pay a fee to enjoy the breathtaking views from the observation platforms brought in enough money to keep the building financially solvent.

For forty years the Empire State Building was the largest in the world. On May 2, 1931, it was dedicated and opened to the public. As thousands thronged through the maze of hallways and gazed in awe out the windows and from the high observation platforms, President Herbert Hoover pushed a button to turn on the lights in the lobby. A soft glow gleamed off the smooth marble in the three-story-high entrance. Then Hoover sent a telegram of congratulation to his old opponent, Alfred E. Smith, the man who had guided the project.

On May 3, tenants began to move in. Some of them are still there today.

Al Smith's dream was finished. It would forevermore jut up into the skyline of New York, becoming the symbol of the city. One of the most admired buildings of its day, it is still the building to visit in New York. The Empire State Building lures more visitors than the Statue of Liberty, Radio City Music Hall, or the United Nations Building. In its great lobby are renditions of seven wonders of the world. They are shown in massive, illuminated panels, giving a three-dimensional picture. Exhibited are the Great Pyramids, the Hanging Gardens of Babylon, the Statue of Zeus, the Temple of Diana, the Lighthouse of Pharos, the Colossus of Rhodes, and the Tomb of King Mausoleus. Then there is an eighth wonder of the world shown in its great panel. It is taller than all the others piled one on top of another. It is the Empire State Building.

Man has by now built larger, taller buildings, and as time passes, he will doubtless outbuild these with newer, even taller buildings. But the Empire State Building will always be the first of the true skyscrapers. It will always be the first of the largest buildings in the world.

And it will forever be one of the most beautiful.

HOOVER DAM

It was dark and somber in the quiet hospital room that night. Visiting the old man in his last hours had been some of the most powerful tycoons in the country, for he was himself president of a great construction empire.

The trouble was, he had not realized his last great dream.

Then a young engineer, Francis T. "Frank" Crowe, moved softly into the room. Company presidents stood aside as he wheeled a tarpaulin-covered table next to the bed. Crowe carefully removed the covering and stepped aside.

The eyes of the old man lighted, and a smile moved across his wan face. He drank in the sight on the table. His hand reached out in an involuntary move to the exhibit.

It was a perfect miniature scale model of a rocky, narrow canyon. The old man recognized it as Black Canyon. Across the canyon model was a jewellike model of a great dam. It was a graceful and statuesque dam, the largest dam in the world in scale model form. The model was perfect in every detail.

The old man on the bed was W. A. Wattis, president of Six Companies, Inc. He smiled up at Crowe. With great emotion he said, "It's a magnificent thing, Frank. It's . . . it's incredible!"

"It's the exact design down to the minutest detail, Mr. Wattis," Frank Crowe said softly. Then he made a promise. "You're seeing it as it will be."

Six Companies, Inc. was the conglomerate of major construction companies brought together by Wattis to build the vast dam, to harness the raging Colorado River once and for all. Wattis did not live to see the dam a reality, but Frank Crowe became the boss of the project and one of the two prime

Black Canyon on the Colorado River, the site of mighty Hoover Dam, before construction began (U.S. Bureau of Reclamation)

movers, along with Walker R. Young of the United States government. For Congress had already approved the construction of the dam over the objections of men as powerful as Norman Chandler, publisher of the *Los Angeles Times* (and owner of nearly a million acres of land in Mexico irrigated by the Colorado River). Other powerful men had lobbied *for* the dam, including another newspaper publisher and a competitor of Chandler's, William Randolph Hearst.

For centuries the Colorado, then an unnamed river, had gently meandered its way south to the sea. But the primeval mountains at the river's source continued to thrust upward and became towering peaks. All of this was before recorded history. Snow, thawing in the spring, added to the river's size, swelling it, increasing its power; eventually it smashed and carved its way, draining more than 242,000 square miles. It formed deep canyons and stunning gorges along the way. By the time the Spaniards came to this part of America, it had become a *giant* of a river.

It carved the Grand Canyon and knifed and ground out other deep canyons along its rushing path. It created Black Canyon.

But could the river be harnessed to provide water for man to irrigate his lands? The California Development Company was founded to channel the water into the vast Salton Sink in Southern California in the early 1900s. The plan worked magnificently. By 1904 more than 10,000 people had moved into the former desert, which had become a lush garden spot of thriving farms. Prosperous cities (El Centro, Imperial, Calexico, Brawley, and others) were developing in what had been renamed the Imperial Valley. It appeared that man had conquered the raging river and forced it to work for him.

But suddenly and without warning the river struck back with terrible floods, and people had to flee for their lives. They were forced to work as hard to rechannel the flow of the Colorado back to its original course as they had worked to change its direction in the first place. Left behind after the great flooding was the Salton Sea, still obvious on all modern maps of Southern California. The river had won again.

In 1905 an engineer by the name of Arthur Powell Davis submitted a new plan to the Bureau of Reclamation for controlling the Colorado once and for all. His plan would force the river to work for man. But Davis's plan involved building a tremendous dam, a dam larger than anybody had ever thought of before. The dam would be built far up the river, hundreds of miles to the north.

Built in either Boulder Canyon or Black Canyon, the dam would hold back and store the water of the Colorado, controlling its flow from season to season. Canals built farther downstream would then safely irrigate the land of the Imperial Valley and land in Arizona. Power from the great dam's hydroelectric plant would provide electricity for future industry and homes all across the Southwest. From the lake formed behind the dam, water could be drawn for millions of people.

It was a grand plan. Perhaps *too* grand.

"It won't work," said prominent engineers.

"It can't be built . . . too big," said others.

"The weight of the dam and the lake created behind it will weaken and crack the earth's crust, possibly causing great earthquakes," warned one respected engineer and geologist.

None of the great European engineers, to whom the Americans looked for guidance, had ever attempted such a dam. If a dam like it ever failed, a flood beyond even Noah's description would wreak havoc all across the southwestern United States.

Still, if it could be built, and if it worked as it was supposed to work, man would have harnessed nature at her worst and put her to work for him. It was a challenging idea to American engineers, who were of a different breed than their European counterparts. They had already been at work measuring and drilling and testing in the suffocating temperatures of the wild Black Canyon region. They wanted to try.

"We won't know if it can be done until then," they said.

The plan was approved, and the six companies banded together under the direction of W. A. Wattis to build the dam. Before he died, Wattis selected Frank Crowe, who worked for one of the companies (Morrison-Knudsen Company of Idaho), as construction superintendent. From the United States government Bureau of Reclamation came Walker Young. Five hundred million dollars had been pooled to pay for the construction.

With the United States deep in the throes of a depression, thousands of jobless and hungry people flocked to the Las Vegas area, where the dam was to be built. On September 17, 1930, Governor Fredrick B. Balzar of Nevada handed a gleaming silver spike to Carl R. Gray, president of the Union Pacific Railroad. The spike had been fashioned from pure silver from one of Nevada's mines. Gray handed the spike to Secretary of the Interior Ray Wilbur, who vigorously drove it into a tie to lock the first rail of the project into place. Then in a speech, Secretary Wilbur named the great dam Hoover Dam after President Herbert Hoover.

Construction was officially underway. Before the end, enough concrete would be poured into the massive structure to pave a highway 16 feet wide from New York City to San Francisco. The dam would stand more than 700 feet tall and hold back a man-made lake 115 miles long and with a shoreline of 550 miles. Where hostile, torrid desert once stretched, a lush Lake Mead would be created and would draw millions of vacationers each year to the former wasteland.

Black Canyon after blasting started. At this point the river is still passing through. (Las Vegas News Bureau)

At first living quarters for the workers were needed. So a complete town was constructed seven miles southwest of the damsite on a high plateau. Homes were built, lawns were seeded, and parks were planted. Streets were carefully laid out and paved. Soon a brand-new modern city was ready, a city with schools, churches, and stores. It still flourishes as Boulder City, Nevada, a beautiful and well-kept town.

During construction of the dam, an intangible idea became very important. America was depression-bound and broke. Americans were frightened and baffled by what had happened to their country. Bread and soup lines forced men to humble themselves for food for their starving families.

But the dam was something to look at with pride. "If we can build a dam *that* big," people were saying, "then things must be getting *better*."

Laborers were earning $4 per day; truck drivers, up to 75¢ per hour. Carpenters and electricians on the dam project earned the same as truckers. Bulldozers and other heavy equipment operators were top men at a princely $1.25 per hour. Even without overtime that came to a whopping $10 per day when thousands of people throughout the country were glad to get $1 or less per day.

Hoover Dam was built of many great concrete forms interlocking with each other to achieve monolithic (one solid mass) design. Each form was five feet deep and sixty or more feet square. A batch plant was built at the site and became the largest cement-providing facility in the world. Day by day workmen would build new forms according to a precise master plan, fill them with concrete, then move on to the next form. Gradually the dam began to take shape as slowly it started to fill the 1,000-foot gap between the walls of stifling Black Canyon. The river had been rerouted during construction.

Great crane-operated eight-cubic-foot buckets of cement would swing into position and drop their loads into the forms as booted workmen stood ankle deep, spreading the mixture. Blast furnace winds bore down through the hostile canyon constantly, sucking the moisture from the concrete and the workers' skins. Although modern records insist that no workers are buried in the cement of the dam, more than ninety men did lose their lives during the construction.

Hoover Dam was the first structure of its type to use the arch-gravity principle of construction, now commonly used in dam construction. This type of dam becomes stronger as more pressure is exerted against it, such as the pressure from the weight of a great body of water behind it.

Construction of the dam progressed day after

were created in 1937. Now skilled and polished guides from the United States Bureau of Reclamation lecture to visitors on interesting inside-the-dam tours. But some people from an earlier day still remember Blackie's gruff descriptions of how it *really* was.

There was Roy Poyser, a resident of Saint Thomas, a community that now lies at the bottom of Lake Mead. Saint Thomas was flooded in the mid-thirties as the waters of the Colorado climbed the upriver side of the new Hoover Dam, seeking a path through Black Canyon. Poyser became known as The Desert Commodore for his early efforts to promote boating safety on Lake Mead, one of the most popular boating resorts in the West. Because the lake is man-made, filling an area of desert and rocky canyon country, there are many deep and clear offshoots and rock-walled, water-filled arroyos to be explored by boat.

Also flooded by Lake Mead were the village of Kaolin, Nevada, submerged at the bottom of the lake, and six and a half miles of the Los Angeles–Salt Lake Railroad track.

The only woman to work at the dam in days past was Ella F. Bell, remembered by most of the nearly 17 million visitors to the dam. During her twenty-six-year tour of duty, she served as cashier and guide. There is a small fee for adults to visit the in-

Many concrete forms interlocked to give the dam its monolithic design. Workers would fill a form, then move on to the next in a precise plan. The photo is shot from the Lake Mead side of the construction. (Las Vegas News Bureau)

The very first concrete forms at the bottom of the dam. Note the hair-raising catwalk from the Nevada side (left) *to the Arizona side and the meandering stairs down to the construction. Both were used frequently by Crowe and any reporters anxious enough for a good story.* (Las Vegas News Bureau)

day, week after week, with concrete pouring into new forms in the great jigsaw puzzle. Inside the dam, new passageways, rooms, and control centers were enclosed as the structure moved ever higher. For the great dam was to do more than merely block the flow of the Colorado River. It would eventually manufacture nearly one and one-half million kilowatts of power with its seventeen great generators.

At the base it was as thick as the length of more than two football fields, 660 feet of solid concrete.

During the six years of construction, the dam workers acquired a mascot, a large dog of non-descript origin named Old Nig. Old Nig didn't belong to anybody, but he was everybody's friend. He knew the dam construction site as well as most of the men. Old Nig would visit the workers' homes in Boulder City, and every day he would carry to the site a lunch in a paper bag, prepared by the wife of the worker at whose home he had spent the previous night. At lunchtim bring his bag and eat with the workers.

Old Nig thought he was a dam work

There are two huge statues on the N of the dam, neither dedicated to the d Created by sculptor Oskar J. W. Hanse designed the visitors' plaza and the showing the exact position of the stars the dam was dedicated, the statues sy spirit of the men who built the dan statues is the grave of Old Nig, still visited by men now elderly who helped dam. The dog was struck by a work killed in 1934.

The dam created other characters struction. Blackie Hardy helped t mighty concrete wall and then becar ficial guide through the many r Blackie, who had grown to love th tourists through free of charge before side of the dam, and Ella sold most of the tickets.

Every statistic on the vast project is awesome. There were 9 million tons of rock cut away and more than 1 million cubic yards of river fill excavated (enough digging to clear a trench 100 feet long, 60 feet wide, and *1 mile deep*).

There was as much structural steel and reinforcing steel as was used in the Empire State Building. The rugged building men who didn't believe that any job was impossible used more than 1,000 miles of steel pipe in the dam. One hundred sixty-five *thousand* railroad cars of sand, gravel, and cobbles, and 900 cars of hydraulic machinery arrived at the site. If all of these materials had been loaded onto a single train, the engine would enter Boulder City, Nevada, just as the caboose was leaving *Kansas City, Missouri.*

Finally, in 1936, the seventy-story wall of concrete plugging Black Canyon and all the accessory spillways (fifty feet in diameter), intake towers, and thirty-foot-tall turbines were finished and installed.

The great dam was ready. More than 4,400,000 cubic yards of concrete had been poured. The American Society of Civil Engineers proclaimed the project one of the "seven modern civil engineering wonders" of the United States. Hoover Dam was the largest structure on earth, larger than the Great Pyramids in Egypt and larger than the greatest buildings ever built by man.

Las Vegas, formerly a sleepy little desert town that was nothing more than a railway stop when a flag was waved, blossomed into a resort area with more than 300,000 permanent residents. Millions of people have visited the town just to visit the dam, more than a half million each year.

The two men most responsible for the dam, Walker Young from the Bureau of Reclamation and Frank T. Crowe of the "Big Six" (Six Companies, Inc.), were with the project from beginning to end. Crowe became a favorite of newspapermen of the day, but only of those reporters brave enough to follow him during his hectic days in the feverish heat and activity from the bottom to the top of steaming Black Canyon. If they could catch him for a few moments, he was always good for a fresh story about the great dam he had grown to love. Where the construction action was the heaviest, they could always find Frank Crowe.

Because Crowe and Young were such a perfect team and worked so well together, the dam was completed two full years ahead of schedule. President Franklin D. Roosevelt, who not that long before had attended the dedication of the Empire State Building as governor of New York, dedicated the project as Hoover Dam on September 30, 1935. The name was soon changed to Boulder Dam, but in 1947 it was again officially changed back to Hoover Dam, the name it bears today.

How long will the great dam last? From inside and outside appearances, from the smooth and polished terrazzo tile floors in the hydroelectric rooms to the extreme cleanliness of the entire area, the dam looks today as it did when it was built. Equipment is constantly being changed and updated, and much of the dam's original generating equipment has been replaced, but the dam itself is the same.

Great intake towers were built on the Lake Mead side of the dam. Water flows into these towers and through turbines to manufacture electricity for the Southwest. (Las Vegas News Bureau)

Hoover Dam completed and in full operation (Las Vegas News Bureau)

There is no proposed life-span to such a project. It would be nearly impossible to erase all traces of the dam, and engineers expect it to stand as it is virtually forever. No natural disaster that would unplug Black Canyon and allow the Colorado River to rage free again can be imagined. The river has been tamed for the use of man. Thousands and thousands of square miles of Southern California and Arizona are now lush and productive farmland because of the dam's control of the river.

Still, young people today have seen many more wonders than their parents did in their entire lives. Is it possible they find the dam tour boring and the dam itself just another construction job?

No! Guides report that the single most used word among young and adult visitors if *amazing!*

The river that couldn't be harnessed waits serenely behind the dam to do its work when called upon, and the dam that couldn't be built is now operating for man.

THE GOLDEN GATE BRIDGE

The beautiful Golden Gate Bridge crosses a gash in the California coast that many engineers felt was impossible to bridge. Joseph Strauss, an engineer and poet, was sure he could span the mile-wide, fast-current opening. (Bethlehem Steel Corporation)

In 1935 a strange thing happened around San Francisco, California. No new bird species had been discovered in years, and ornithologists were excited about an unusually marked sea gull found in the area. They hurried from around the world, hoping to see, or better yet to photograph, one of the rare new species of sea gull.

The bird appeared much like any other sea gull, with white and gray coloring, and graceful, sweeping wings. But this species had a red head. The discovery was astonishing. It was one thing to find a new bird with a slightly different eye, or a tinge of feather difference, or even a slightly different shape. But a *red head?* A complete change in an established species?

Scientists were astounded.

Eventually the bird was catalogued and even depicted in publications of the day. It was accepted and studied.

Finally the truth came out. The birds were not a

new species at all, but merely hungry members of the standard old sea gull family. Workmen painting the staggering array of steel on the new Golden Gate Bridge under construction took their lunch hour far above the ocean, with legs dangling from girders. The seagulls knew about lunchtime and came flying by or landed to beg food from the painters.

So the sport among the painters was to see who could most adeptly dab a blob of "international orange" paint on the head of a passing sea gull. The gulls were willing to take the risk for the food being offered, and a new species was almost born as landlubbers far below and finally scientists noticed the red-headed seagulls.

There is a huge inlet in the San Francisco area leading to a giant bay. To go directly across the water is plainly the most efficient way to travel north up the coast. But travelers needed a ferryboat before 1937.

The space between was much too large for a bridge, it was generally conceded. Still, a bridge would be the ideal solution to a knotty transportation problem that had plagued residents and travelers for years.

The Indians who once lived in the area had an explanation for the formation of the unbridgeable gate. Once, according to legend, the area was a fertile valley separated from the ocean by a range of mountains. The great sun-god favored the valley and often paused there to visit on his way from east to west. Once while visiting, he fell in love with a beautiful Indian princess.

Brashly attempting to kidnap her, he lifted her high and begun to run away to the west. But in his haste he stumbled and began to fall. The poor princess slipped from his grasp and fell back to earth.

As the sun-god fell to the ground in grief at the loss of his princess, his mighty arm smashed into the mountain range between the valley and the ocean. A great cleft was broken open. The ocean waters rushed in to fill the valley, and the bay was created, remaining there to this day.

Deeply saddened at the loss of his love, the sun-god gently placed the body of the princess above the water atop Mount Tamalpais. The outline of her form can still be seen resting there.

Geologists have other explanations for the formation of San Francisco Bay, first discovered by Sir Francis Drake and later named by the American explorer John C. Frémont.

Through the inlet rush the waters of the Pacific Ocean. The mile-wide gap in the coastline was one of the most rugged on the Pacific. For generations the great engineers felt that bridging the gap was impossible.

Joseph B. Strauss was a wisp of a man physically. He was a dreamer, a writer and poet with many published poems to his credit. He was also a giant among engineers, especially bridge engineers. He had built the great Republican Bridge at Leningrad, in Russia; the Longview, Washington, bridge over the Columbia River; the beautiful Arlington Memorial Bridge in Washington, D.C., and several hundred other great bridges.

Strauss was a bridge builder extraordinary, but the one span he wanted to cross had always eluded him. A lover of the great trees in Northern California and also of the area around San Francisco, he had studied the Golden Gate and yearned to build across it.

Strauss felt that he could do the job; but San Francisco didn't think the bridge could be built. Ferryboats had served for years and could continue to serve, slow as they were. Strauss studied the rugged gash in the Pacific coastline.

Finally in 1929, after years of deliberation, authorization was granted to construct a massive bridge over the Golden Gate. Joseph Strauss felt great joy in his heart when he was appointed chief engineer of the project. It was the opportunity he had awaited: the chance to build the impossible bridge.

It would be a suspension bridge, Strauss knew, with the roadway literally hanging from monster cables that were supported by two great towers and deep anchors embedded into bedrock at each end. It would be a sweeping, beautiful bridge, a great monument as well as a transportation device. It would stretch over the fast-moving waters between the Presidio of San Francisco on the south and the Fort Baker Military Reservation on the north, where the bay narrows to a few feet less than one mile in width. The bridge itself, from abutment to abutment, would be 8,450 feet long, and the support towers would rise above the Golden Gate higher than sixty-five-story skyscrapers. The roadway would be suspended from cables nineteen stories above the water.

Still, there were many political matters to settle before the bridge could be built. On September 8, 1930, Secretary of War Patrick Hurley approved plans for the Golden Gate Bridge as designed by Strauss. The plans called for a San Francisco pier supporting the south tower and a Marin pier for the north tower. The north tower pier would present no special problems, Strauss knew, since it was connected to the rocky land near Lime Point Lighthouse. But the south tower was different.

Enclosed by a great fender to protect it from wayward ships and fast-moving tides, the south pier was to be built far into the bay, more than 1,000 feet from shore. To be built properly, it would require a long trestle from the shore near Fort Point.

The trestle was eventually built and served as a supply roadway to the south pier, but twice it was severely damaged. Once an out-of-control ship slammed into it, and once it was partially wrecked

Strauss (left) *and his foremen were everywhere on the bridge construction, even on the tip of the towers where the cable housings were placed.* (San Francisco Public Library)

by a wild storm. Both times construction of the south pier was delayed while the supply trestle was repaired.

Even before construction began, there were more politics to be overcome by the impatient Strauss. In November 1930, a bond issue to pay for the bridge was passed by San Francisco voters; 145,657 voted yes, and 46,954 voted no.

Today there are more people than that in any one of the many districts of San Francisco, partly as a result of the beautiful bridge.

It was not until January 1933 that actual construction of the piers and the anchors began. First the piers and anchors, then the giant towers, and finally the cables to hold up the roadway were to be built. The towers were the tough part, the cable stringing the exciting part, and the roadways the satisfying part.

To the artistic Joseph Strauss, who wrote a final lovely poem as an ode to the bridge he built, it was all a joy. He was everywhere on the project, underwater, high in the towers, crossing the cables on tiny work platforms, and back in his office to correct problems and polish details. The magnificent bridge became Strauss's greatest poem.

Slowly work on the south tower progressed. Meanwhile, the north tower was nearly completed and ready for the next step. Lime Point Lighthouse, a complex that once dwarfed the beginnings of the north tower, was by then a dwarf itself under the towering north structure. The towers reach from 110 feet below the surface to 746 feet above it.

Inspired, Strauss worked on at his great dream. Detractors were still hurling insults and making dire predictions, though they were lessening as the bridge grew larger.

"In the case of war, one shell could drop the bridge and block one of the most strategic harbors in the world!" some insisted.

"It will never withstand an earthquake," said others. It has, of course, many, many times.

Finally work has completed on the south tower and the anchorages at each end of the bridge. The north tower was ready and waiting. Cable spinning could begin. There would be two main cables made up of many, many smaller wires. Each main cable would support one side of the roadway through a series of hanging cables, in typical suspension construction.

But these main cables would be over *1 yard thick,* the largest ever built, and *7,650 feet long.* They would be spun by a system developed by John Roebling for the Brooklyn Bridge in the 1860s. Each huge cable was eventually made up of 27,572 individual cables for a total cable length in the entire bridge of more than *80,000 miles.*

The cable spinning began with workmen scrambling up along dangerous one-man catwalks at dizzying heights above the water. Below, six and one-half knot ice-cold tides swept back and forth, the same tides that for so many years were the only thing to prevent escapes from nearby Alcatraz Island. The very first federal prisoners on The Rock, as the island became known, watched the great bridge being built from the time that the prison opened in 1933 to the completion of the bridge in 1937.

At right, above: *Before cable spinning could begin, workmen had to install a high catwalk from anchorage to anchorage between the two great towers.* (San Francisco Public Library) Below: *Each main cable, shown just above the workmen's heads, was to be over 1 yard thick. The structure riding on the cables is the spinning machine, bringing still another weblike cable to join the whole.* (Bethlehem Steel Corporation)

Before each 11,000-ton cable was finished, the spinning machines had gone back and forth hundreds of times, laying out cables in a precise way to distribute the great weight of the roadway and traffic.

Workmen were provided with as many safety precautions as possible, but high iron men come to trust their instincts more than they trust safety devices. They scamper about on ironwork at heights that would make most people faint. They lean and work and jump from spot to spot with frightening aplomb.

Still, a great net was stretched under the cables, under the position of the roadway, to catch any falling workers. And it *did*. The net saved a total of nineteen falling men, men who have since then been members of an exclusive "close call" club formed during bridge construction. But thirteen other men died on the project, ten in one day when a heavy steel girder jumped free. The girder

The framework for the roadway progressed in both directions from each tower and suspended from hangers at the ends of the cables. The access trestle (bottom left) *and the high catwalk* (under the main cables) *were still in place when this photo was taken, but they have long since been removed.* (Bethlehem Steel Corporation)

knocked the men off the bridgework, then carried away their last hope, the safety net, as it fell ahead of them to the sea far, far below. The other three workers were killed much later, in 1954, in a similar accident during a project to stiffen the center span of the bridge with extra trusses.

From the high catwalk and with the use of the spinning machines, the main cables were finally completed, squeezed perfectly round by a hydraulic press, and then sheathed with a protective covering to prevent rust. The long sets of roadway support cables (each made up of seven individual strands, with each strand made up of more than two dozen smaller cables) were hung from the main cables. At the end of each of these was a hanger to be fastened to the steel framework for the road surface.

Meanwhile, the bridge approaches were being completed with rerouting of roads and streets, and the toll plaza work was being finished.

By then the bridge arched high over old Fort Point on the southern shore. The fort was left intact to give visitors an idea of the early defense of San Francisco Bay. The fort can still be visited today, and the bridge viewed from a fine vista point nearby.

From the main cable finally hung a long series of roadway suspension cables (now being replaced, one by one. Four-inch sections of the original cable sell for from $25 to $50 to collectors and bridge buffs). The framework for the roadway was started. Meanwhile, workmen were dismantling the catwalk from its place far above, and other workers were installing aircraft beacon markers atop each of the towers.

Work on the roadway progressed outward and shoreward from each tower to meet finally at the center point of the mid-span and to reach both shores. Then the roadway was added, and the grand bridge was ready. Strauss looked proudly at

STEEL COMPANY

Finally the framework was complete, and the roadway could be added to finish the great bridge. (Bethlehem Steel Corporation)

his achievement, a monument that would stand to his design genius and his construction ability.

Then, typically, he wrote his poem. The opening stanza said:

> At last the mighty task is done;
> Resplendent in the western sun
> The Bridge looms mountain high;
> Its titan piers grip ocean floor,
> Its great steel arms link shore with shore,
> Its towers pierce the sky.

Opened to foot traffic on May 27, 1937, it was finally opened to a stream of autos the next day. The long stream has never stopped since.

The Golden Gate Bridge is not a rock-solid monument. Rather, it is designed to give with the elements, to sway with the wind. It will likely stand until people decide to remove it.

If one side span of the bridge is loaded with "live" weight (auto and truck traffic, bumper to bumper) and the center span is empty, the top of the nearest tower will lean exactly twenty-two inches toward the loaded shore span. That is, if the temperature is 25 degrees. The lean changes slightly as the temperature changes, and engineers like Strauss knew exactly how much to allow for.

Say the temperature is 105 degrees, possible in that area in midsummer. If one end of the center span is loaded with live weight and the nearest side span is empty, that supporting tower will deflect eighteen inches toward the center. As the temperature changes, the lean will change.

A 40-degree rise in temperature from the normal of 65 degrees will cause a downward movement of the center span, when fully loaded, of ten feet. If the temperature is 40 degrees lower than normal, and the center span is empty, it will rise ten feet at mid-span.

This gives the center span a total extreme range in rise and fall due to temperature and weight—not wind—of up to twenty feet.

A strong wind can actually deflect the great towers to the *side* a few inches, and if the bridge should be subjected to a wind speed of 120 miles per hour (it never has been), the center span will swing out one way or the other, depending on the direction of the wind, up to twenty-one feet. That gives the bridge's center span a total sway possibility of forty-two feet from side to side.

The Golden Gate Bridge is truly a flexible monument.

And it is never completely finished. Every day visitors can see painters working their way across the structure of the bridge. Fifteen thousand gallons of paint are used every year in this never-ending painting job. When the painters have worked all the way across the bridge, they go back to the other end and start all over again.

Inspectors rove over the bridge all the time—inside and to the top of the towers, under the framework of the roadway, and on the roadway surface.

New cables are strung as needed, and measurements are constantly taken to make sure the bridge remains sound. The people who work on the bridge feel that it is a living thing, not just a dead structure of steel and cable.

In the center of the center span is a foghorn, the only one in the world that *beckons* ships rather than warning them away. The horn is really two horns called diaphones, blowing together. They guide captains of great ships to the exact center of the channel, the deepest point. There is no ship in the world that cannot pass under the Golden Gate Bridge, and recently a mighty floating oil drilling rig was towed under the bridge (with scant *inches* to spare).

A full staff of people manages and operates the bridge. There is a fire department for auto fires on the bridge, and a service and tow department (*no charge* except for gasoline at standard prices, if you run out). There are even crews to pick up any stray dogs or cats that somehow manage to get on the bridge.

The mighty Golden Gate Bridge is a beautiful addition to one of the most beautiful parts of California, next to one of the most beautiful cities, in one of the most beautiful parts of the world.

THE GATEWAY ARCH

Although the area was little more than a rustic trading post under the prosperous hand of Frenchman Pierre Laclede, it had become the necessary last stop before heading into the western wilderness. Beyond Laclede's little store and cluster of outbuildings on a grassy mound at the edge of the mighty Mississippi River was nature at her worst. Beyond were rugged mountains, man-killing deserts, and plains as far as the eye could see or people could travel in many days. Wild animals waited out there.

Indians also waited, earlier Americans who resented the white man's attempt to encroach on their property and were willing to fight to stop him.

In 1803 three men signed a piece of paper called the Louisiana Purchase Treaty. In one sweep of his pen, Napoleon's representative, the Marquis de Barbé-Marbois, doubled the size of the United States as President Jefferson's negotiators, Robert R. Livingston and James Monroe, watched. The French wilderness became the American wilderness.

Laclede's fur trading post, by then called Saint Louis, suddenly became the last truly civilized stop on one of the greatest migrations ever known. Pioneers, anxious to settle in the vast new land, stopped at Laclede's for final stores for their dangerous wagon train journey. Fur traders stopped there, and so did adventurers, scientists, and explorers. Missionaries made Laclede's post the jumping-off spot on their quest for new converts among the Indians.

Saint Louis was the last of what was considered settled land (two or more persons per square mile). As such it became the gateway to the Wild West. From 1803 until 1890, when the land west of the Mississippi became so populated that there was no longer a frontier line at Saint Louis, the river settlement continued to grow by leaps and bounds.

To commemorate the unique position of Saint

The mighty Gateway Arch in Saint Louis, shot from across the Mississippi River, with the Saint Louis skyline in the background (Bi-State Development Agency)

Louis in the settlement of the western part of the United States, citizens of the city formed a group known as the Jefferson National Expansion Memorial Association in 1934. They studied old books and papers and finally found the exact location of Pierre Laclede's trading post, by then along a riverbank levee in the midst of blocks of decaying and largely unused business buildings. The group bought the entire historical area and then had it cleared.

But then what? A few statues? A park? How could they best mark the spot where the sodbusters, traders, explorers, and cattlemen had crossed the Mississippi to open the vast American western wilderness? It had to be something special, something very special.

Congress made the area a National Historic Site, and with the federal government's help several plans were considered. They included using the land for fancy groups of buildings, museums, statues, and other typical monuments to the Louisiana Purchase. But nothing seemed just right to the people of Saint Louis.

Then World War II came along, and the land was idle, serving as a series of parking lots and a catchall area for city equipment storage.

The problem was, nothing really special had been found. The citizens of Saint Louis weren't happy with statues and museums. In 1947 the Memorial Association decided on a national architectural competition with prizes large enough to attract the top designers in the country. New ideas flooded into Saint Louis; among them were some very beautiful plans for the waiting land where Laclede once worked.

In 1948 the grand prize winner of the competition was announced. He was a young architect of Finnish descent, Eero Saarinen. His design was a magnificent upward-sweeping stainless steel arch, facing west and standing alone and uncluttered. It was a stunningly beautiful memorial to the gateway city.

The arch was startling in its simplicity but complex in design. It would call for the development of many new construction techniques. But builders were sure it could be done.

Under the great arch, underground and out of sight, was to be a Museum of Westward Expansion. Visitors would see exhibits on how the West was won and on the men, women, and children who were victorious. But the most exciting part of the monument was the fact that visitors could ride to the top. From windows more than 600 feet above the ground they could look out on a vast panorama of Saint Louis, the Mississippi River, and the first many miles traveled by the wagon trains so many years before.

Saarinen and his associates planned to follow the construction from beginning to end, and the associates did so. Saarinen himself died before the first huge triangular base section was set in place. He never saw his greatest dream complete, but he knew it was going to be done.

Saarinen's design was in the shape of an inverted, weighted catenary arch, a classic shape in architecture. It is the shape a chain assumes when hung between two points. It is the strongest arch of all. All the stress of the weight of the arch is transferred to the legs and foundation. A weighted arch is one with larger, heavier legs, one that is tapered toward the top.

One construction problem was apparent as ground was broken in 1962. Already the museum shell had been built and covered with earth, and the footings for the arch had been sunk sixty feet into bedrock. It would be simple enough to raise preformed triangular sections one on top of the next from large cranes on the ground.

But what would happen when ground cranes could no longer reach up high enough on the legs to place new sections? And what would happen to the legs as they began to lean in toward each other? Six hundred thirty feet apart at the base, the legs would have to meet six hundred thirty feet in the air.

The match would have to be *perfect*. A deviation of only 1/64 inch in the location, size, positioning, or angling of the huge fifty-four-foot base sections would mean that the inward-leaning legs would completely miss at the top.

Measurements had to be *exact*.

The great triangular base sections were placed, and the legs began to climb into the clear Saint Louis sky. They were two triangles, one inside the other with a space between. The inside wall was standard steel plate, welded by skilled construction men. The outside wall was shiny stainless steel. Great care had to be taken to protect this gleaming finish.

Concrete was poured between the walls of the legs up to the 300-foot elevation. The concrete and prestressed steel rods gave the bottom of each leg strength and solidarity. Engineers calculated that the top of the finished arch would sway only eighteen inches, not enough for visitors in the high observation room to notice, and then only if the winds reached an unheard of 155 miles per hour.

The arch on the shore of the Mississippi is immensely strong and durable.

Gradually the legs rose, with ground cranes lifting section after section into place. As each section was located, welders began their work. Then concrete was poured to lock the new sections with the earlier ones. Meanwhile, in the hollow centers of each leg, stairs were built, and equipment was put in place for the elevators and unusual self-leveling passenger trains.

Finally, the day came when the ground cranes could no longer reach the tops of the legs. But engineers were ready. Unique climbing cranes had

The climbing cranes have reached a point at which the legs have to be braced apart. (Jefferson National Expansion Memorial)

been developed. Tracks had been laid up the outside of each leg, and a self-leveling crane was placed on a platform on each track. From that day on, the cranes climbed up and in toward each other, lifting sections, setting them in place, then crawling across them to lift the next sections. Care was taken to protect the shiny stainless steel outside skin of the arch from the cranes.

Each huge derrick platform was self-contained. Once up, workmen didn't have to come down all day long, for each platform was a complete work station, with warm rooms for coffee breaks, rest rooms, and emergency first-aid facilities.

Up went the cranes and the workers, every day climbing higher and higher. As the legs progressed upward, they also began to lean in toward each other more and more.

The more the legs leaned inward, the greater became the danger of workmen slipping and falling. But a marvelous safety record was also being set.

Not a single workman had fallen, nor had any serious on-the-job injury been recorded. Nobody thought the record could be maintained, for on every major construction job there is injury and often loss of life, but as the work went on, the record continued.

Though workmen were finally clinging to the slippery outside of the monument more than 600 feet in the air, not a single fatality was recorded nor a single major injury treated. During the entire construction of The Gateway Arch, from groundbreaking to dedication, not a single life was lost.

From the 25,000 tons of concrete and steel rods anchored deep in bedrock, the gleaming legs of the arch rose ever upward. They could be seen for miles, reaching toward each other. Above 300 feet, the space between the inside and outside walls of each leg was braced with cross members, but no more concrete was poured. A few drops of water from far north, from the source waters of both the Mississippi and the Missiouri rivers, had been added to the final concrete mix.

As the legs climbed, they began to taper from their ground measurement of fifty-four feet per side. Gradually and gracefully they were closing down to an outside measurement of seventeen feet eight inches at the peak of the arch.

Plying back and forth on the river far below were steamers carrying cargo and old stern-wheeler riverboats to remind workmen of the historic past of Saint Louis. On the decks of the vessels, crew and passengers gaped up at the great curving legs.

Soon workmen were shouting across the nearly horizontal gap between the two legs. A framework of steel had been erected between the legs at the 548-foot level. This 58-ton temporary stabilizing strut braced the two legs apart as the curve inward grew extreme. Engineers had planned that the two legs would begin to lean inward by their own tremendous weight. The strut would hold them apart

With the temporary stabilizing strut in place, the climbing cranes could work on. Soon men were able to shout at each other across the gap closing between the legs. (Jefferson National Expansion Memorial)

at the correct distance during erection of the topmost sections.

Far below, railroad car after railroad car stopped. On each flatcar was another section, carefully measured and precisely built to fit exactly where it was supposed to. The sections were built in Pennsylvania and shipped to the construction site in Missouri. Thousands of people along the route of the railroad had seen and wondered about the huge triangular double-walled sections heading toward Saint Louis.

The tips of the two legs stretched almost horizontally toward the crest, and a dramatic juncture drew near. On the railroad cars below, the sections now had windows in them. These were destined to be the long observation room at the very top of the arch. Now workmen could easily speak across the gap without shouting. A huge safety net was stretched across below the opening, just in case. It was never used.

Finally the great arch was "topped out" with the final windowed section, joining the two legs and completing the observation room at the top. (Jefferson National Expansion Memorial)

The creeping cranes began to lift a strange new load. Large electrical generators to power the passenger trains inside the legs were set in place under the floor of the observation room.

Finally, on October 28, 1965, the day arrived. According to plans drafted by Saarinen many years before, the two gigantic legs were exactly two and one-half feet apart. Imagine the thrill of stepping across that gap in the floor of the observation room, as many workmen did during preparations for the final closing. On the ground was an eight-foot section exactly as planned. This section had been built to fill the final space.

But first, huge hydraulic jacks were set in motion to pry apart the two 20,000-ton legs. Gradually they were inched apart until the two and one-half foot gap became three, then four and five feet. Workmen and engineers watched carefully as the gap increased. Stupendous forces were at work, and precision beyond imagination was required.

Finally, the gap was wide enough, and the last section was slowly lifted. Thousands watched as it was poised above the opening. It was early in the morning, before the heat of the sun could expand any of the metal in the arch. The work went slowly and carefully. Too slowly. At noon engineers called for the Saint Louis fire department to spray down the southern face of the arch. The sun beating upon it had expanded it eleven inches more than the shady northern face. Workmen waited as the metal cooled. Then the final section was finally lowered into place.

The fit was *perfect*. The amazing engineering measurements, until that instant not proved, had been absolutely precise. Eero Saarinen would have been pleased, for his arch, the arch he never saw in reality, was closed and complete. The Washington Monument had taken forty years to complete, the Saint Louis Gateway Arch only three; yet it is higher and much more complex.

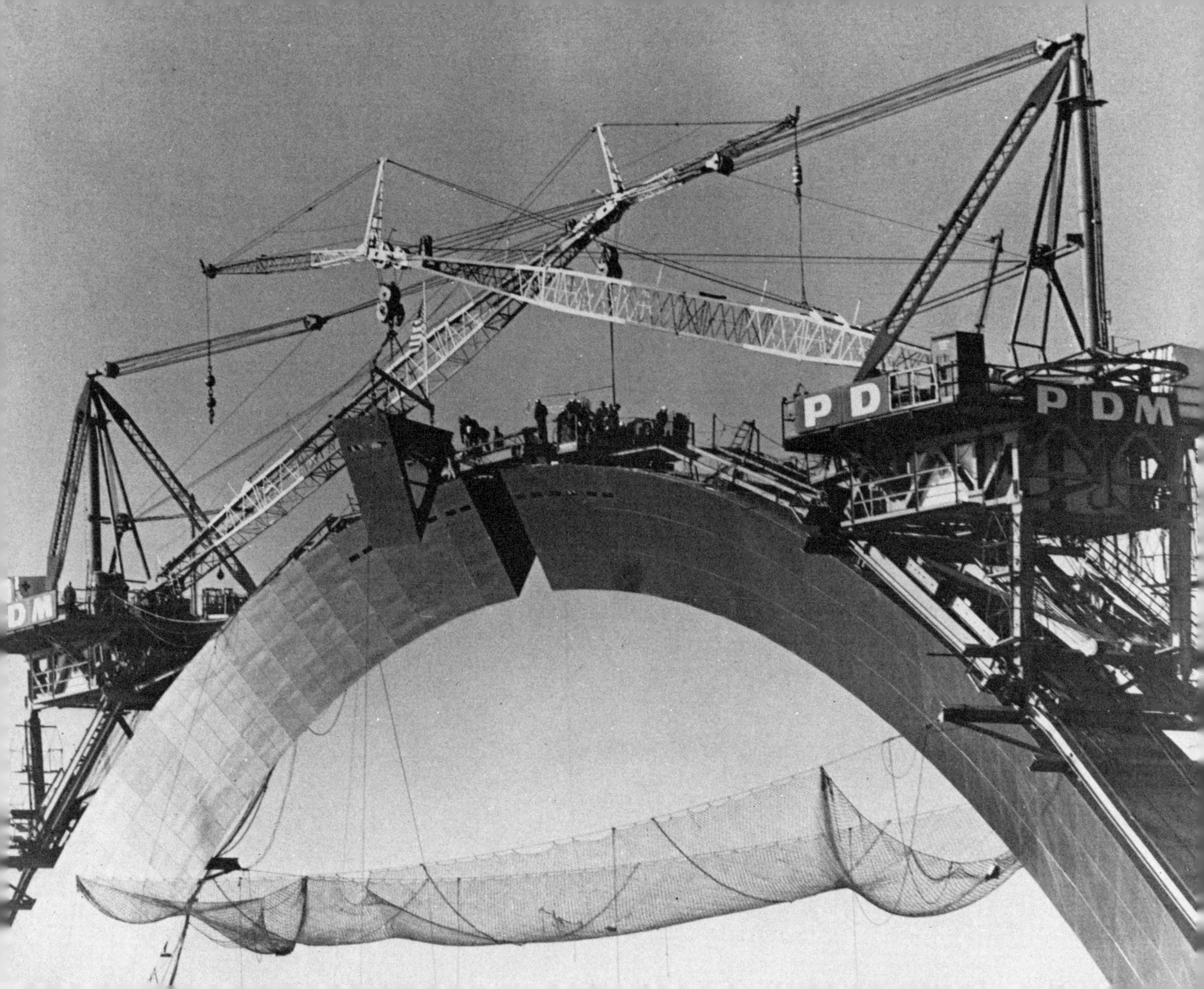
PDM

Workmen bolted and welded the final section in place; then the huge jacks were released. The great weight of the legs clamped against the final section, and the arch was whole.

But not yet ready for visitors.

The huge strut was removed, and the derricks that had crawled all the way to the top began to creep down along the gleaming legs. As they moved down, workmen removed the tracks from behind them. Surface holes in the shining stainless steel where bolt heads had been fastened were filled and ground smooth. Finally, as the workers moved down, the outside skin was cleaned and polished.

For thirty miles in all directions, the arch began to shine in the sun as it does today, for stainless steel is almost ageless in its beauty.

Inside the hollow legs, more workmen were installing elevators, finishing the observation room, and preparing the self-leveling passenger trains that would carry sightseers to the top through each leg. Visitors were to board the eight-car, forty-passenger trains from the underground museum. Seats in the cars were at the correct angle for boarding and comfort; then as the trains started upward, the cars were to swing so the seats would always be level. Finally the cars would stop; passengers would disembark, walk up a short flight of steps, and enter the sixty-five-foot observation room.

From there they would look out the thirty-two narrow windows, as visitors do today by the thousands. For the inner railway and the vast underground museum were opened to the public as a National Historic Monument in 1967.

The great Gateway Arch is more than twice as high as the Statue of Liberty. It is higher than the United Nations Building. It is the highest national monument of all. From its sixty-story-high observation room can be seen old and new Saint Louis;

Illinois across the river; and the great sweeping curve of the Mississippi itself, where Mark Twain once steamed and where pioneers once heard the stirring call "Wagons *hooooooo!*"

No, Eero Saarinen never saw his final great dream come true. But the awesome steel arch soaring high over the spot where Pierre Laclede sent thousands of pioneers on their way into the wilderness will remain a tribute to Saarinen's magnificent design and building skills for ages to come.

The sixty-five-foot observation room at the very peak of The Gateway Arch. Trains bring visitors to the top. Over each window is a map of what can be seen to the east or west. (Bi-State Development Agency)

THE ASTRODOME

It was probably the most peculiar groundbreaking in the history of construction. There they were—construction people, county supervisors, and other officials—on a windswept Texas prairie. They were standing in a typical pose and smiling self-consciously at the cameras.

But where were the traditional silver shovels?

These men were all armed with western-style Colt .45s. They looked like a band of modern robbers waiting for the bank to open or the train to come by. In suits and ten-gallon hats, they awaited the word.

Then the signal came, and they all drew their weapons and began blasting away at the ground to "turn" the first earth for the great construction project.

Unusual, and yet not so unusual when you consider what these Texans planned to build. They were going to construct a football-field-sized sports arena and then put a *roof* on it. It had never been done before. Some architects and builders felt that the problems of covering an entire sports field and huge grandstands, the largest area ever to be covered in history, might be too great.

But the rewards in comfort for spectators and players and the perfect control of weather made the project too tempting to resist. Nobody on that day had any idea that one day in the future it would rain *inside* the great dome or that for the first time in sports history a game would be rained out even though the seats and field were perfectly dry and ready.

Texas, in the early 1960s, was a state without a big-league baseball team. But what the state *did* have was a sports enthusiast and man who had the ability to make things happen. His name was Judge Roy M. Hofheinz.

Hofheinz was on a trip to Rome, Italy, when he thought of the idea of a domed stadium. Gazing down on the Colosseum, he heard the guide say

At an unusual ceremony, members of the board of the Colt .45s baseball club (later renamed the Astros) as well as other officials broke ground with pistols instead of shovels. (Houston Sports Association)

that once in the dim past the great arena had had a velarium. This was a device that was pulled over the amphitheater when the weather was unsatisfactory. Why not such an arena in Texas? thought Judge Hofheinz. Anything the Romans could do, the Texans could do better, and bigger for that matter. The judge sketched his idea on the back of an envelope and stuck it in his pocket.

Back home, Hofheinz discussed the idea with architect Buckminster Fuller, famous for his geodesic domes.

"Fuller convinced me that it was possible to cover any size space if you didn't run out of money," Judge Hofheinz said later.

But what about a space as big as a football field and baseball field combined? Not only that, but a space that could be *converted* from one field to the other? It would mean moving thousands and thousands of grandstand seats in a relatively short time.

Roy Hofheinz was accustomed to getting what he wanted. A lawyer at nineteen; a member of the Texas legislature at twenty-two; a county judge of Harris County for four terms, beginning when he was twenty-four; and Houston's mayor for two terms, beginning when he was forty; he was a man who could sway people. Hofheinz formed the Houston Sports Association with George Kirksey, Craig Cullinan, Jr., and Bob Smith, and then went off to obtain a sports franchise.

Carrying along a model of the proposed domed stadium, the group brought back to Texas a franchise for the Colt .45s, the first major-league baseball team in the South.

"I knew that with our heat, humidity, and rain, the best chance of success was in the direction of a weatherproof, all-purpose stadium," explained Hofheinz, "a spectators' paradise that could be used for other sports as well."

Hofheinz was the type of man who would have been amused had he been able to see what was

On pages 106 and 107: *Progressive construction phases of the Astrodome* (Houston Sports Association)

going to happen once in the future. Just so it was *only* once.

The Houston Astros were scheduled to play the Pittsburgh Pirates in a baseball game on June 15, 1976. The teams were there, but umpires and fans were unable to reach the domed stadium because of heavy rains and flooding *outside*.

So, finally, the game was postponed, and both teams sat down to dinner at tables set up on the field. Along with twenty workers and random fans who had managed to get to the field before the heaviest of the flooding, the teams enjoyed dinner while the rain beat on the roof far above.

"It's not a rain-out," said one official. "It's a rain-*in*."

Three years passed between the granting of the franchise for the team in 1960 (the groundbreaking with bullets was in 1962) and the beginning of construction in 1963. The stadium was an impossibility in the opinion of many Texans. It would cost far too much to make it feasible, and besides, the roof would probably fall in before it was ever finished. There had never been a domed football-baseball field before, and the first one wasn't going to be in Texas.

These folks didn't reckon with Judge Roy Hofheinz. He pushed through bond issues, and long after the Colt .45s began their first season, the great roofed stadium was finally started. Eight hundred workers reported to the site on March 18, 1963.

Gradually the stadium to be called the Astrodome took shape on the Texas plain outside Houston. It was going to be a magnificent structure. While the steel climbed on the outside and workers scrambled about on the high roof, other workers were laying tracks and building grandstands inside. Judge Hofheinz, who had a luxury suite built in the stadium for himself, was on top of every part of the job. When a decision was needed,

November 1963

April 1964

July 1964

October 1964

Hofheinz would make it and be responsible for it.

The stadium structure covers 9 1/2 acres of land. The stadium has an outside diameter of 710 feet. The floor of the stadium is 25 feet below ground level. The roof is 208 feet tall. Inside are 41,000 cushioned theater-type seats and 4,000 extra pavilion seats for baseball. For football there are 52,000 seats, for conventions 60,000 seats, and for boxing 66,000 seats.

An *eighteen-story building* could be built at second base and not touch the roof. A two million dollar scoreboard, the real personality of the Astrodome, was constructed.

Into the stadium was built a 6,000-ton air conditioner that circulates 2 1/2 million cubic feet of air every minute. Every minute 250,000 cubic feet of fresh air are drawn in and conditioned by electronic filters and activated charcoal odor removers. Smoke and hot air are drawn out through the roof.

Engineers on the project were not worried about the roof or walls caving in, as large as they were. But they did worry about something else. There was a period during construction when the walls were up and the roof was being finished, but the overall structure was not yet anchored solidly together.

There were several hurricane warnings during this phase of construction, and engineers were worried about the possibility of the roof of the Astrodome becoming a great Frisbee disc. The shape of the roof gave it a certain amount of lift. Wind passing over the top of the roof could create the same effect as wind passing over an airplane wing, creating a low-pressure zone over the roof. The roof could have lifted up to fill the void, just as an airplane wing lifts, and gone sailing out over the Texas prairie.

Now the roof of the Astrodome—finished, enclosed, and solidly attached to the walls—can withstand winds up to about 135 miles per hour, with gusts up to about 160 miles per hour, before problems might begin to develop.

There was consideration given to staffing the dome with a crew of pigeon hawks, except that the hawks would probably have become just as much a problem as the early "Astrodome Air Force" became to workers. Somehow a flock of pigeons got inside the dome during construction, and as the structure was enclosed, the pigeons were trapped. They would rise majestically from the strawberry-colored seats, flapping their wings furiously and heading for the sunlight. But then, bewildered, they would back off and finally land in the high girders of the ceiling. Then off they would fly again, seeking a new route to freedom.

Meanwhile, far below, disgruntled stadium workers would follow along, cleaning the soiled upholstery of seats and grumbling about nature's only indoor air force. The pigeons are still there.

Originally a beautiful natural grass playing field was built into the Astrodome. To help the grass grow, the roof was covered with clear plastic panels so that the ultraviolet rays of the sun could pass through. But the sun became the horror of stadium engineers and baseball and football players.

The grass grew into a lush carpet, but as it grew, it released moisture into the air of the stadium. It became possible for clouds to develop *under* the dome and for rain to fall *inside* the stadium. The huge air-conditioning system was set at its highest level of power just to keep the atmosphere dry.

Meanwhile, fielders and pass receivers were having problems with high balls getting lost in the glare from the roof.

So finally the roof was coated to cut down the glare, and the grass was replaced with an artificial grass, the world's first Astroturf (now used in hundreds of sports installations around the country).

The panels in the roof allowed light in to help the grass grow, but the grass was replaced with Astroturf. Then the panels were coated to cut down the glare—a problem for players trying to catch high balls. (Houston Sports Association)

On April 9, 1965, the Astrodome opened at last. Thousands flocked to the new domed stadium, a wonderland of beauty, comfort, and technical accomplishments. It was the world's largest indoor arena, the world's largest air-conditioned building, and it enclosed the largest clear span (no interior columns or supports) of any building ever built. It was equipped with the world's largest scoreboard and the world's most pampered grass, (eventually to be changed to Astroturf).

It was a Texas-sized building. Women in silks and furs, men in tuxedos, and the president of the United States, Texan Lyndon B. Johnson, watched as the Astros (the new name for the Colt .45s) beat the New York Yankees 2–1. Far up in his luxurious suite, Judge Roy Hofheinz watched with glowing satisfaction. He had accomplished the job many said was impossible.

Below, Mickey Mantle hit the Astrodome's first home run. He hit it during the first air-conditioned baseball game in history. It was a night of firsts, and fans screamed their approval of everything . . . including the actions of the batboy.

Never without words, a beaming Judge Hofheinz, explaining the new name for the team, said, "It is in keeping with the space age in the capital of the space world. And we think that the domed stadium makes Houston the capital of the space world." The judge's eyes were twinkling.

In only four months, attendance for Astro games at the Astrodome exceeded *two million*. More than fifty thousand people packed in for the Houston Boy Scout Circus. Ringling Bros. and Barnum & Bailey, another great circus, came to the Astrodome and smashed all previous attendance records. Weather was no longer a problem. No longer would the circus, at least in Houston, have to worry about washed-out midways and muddy rings.

In September 1965 they pushed the magic buttons for the first time and changed the baseball field to a football field. For on the eleventh, the University of Houston was scheduled to play Tulsa in the first football game ever to be played indoors.

Changing the Astrodome from a baseball to a football stadium is not an easy task, but considering what is being accomplished, it is not that difficult. Strips of Astroturf are removed and rolled up; then 10 inches of top soil are removed from over the rails. Redwood boxes that protect the rails are lifted, and the tracks are cleared. Then a button activates a 10-horsepower motor that slowly moves two 5,000 seat sections from the sides of the baseball diamond to parallel positions on each side of the football field. The seats move 35 degrees, or about 140 feet.

Then the tracks are covered with their boxes, the soil is replaced, and Astroturf covers the part of the field that is exposed. The pitcher's mound is skimmed down flat, the football field is marked, and the stadium is ready for the new game. Extra seats can be added in each end zone area if tickets sales warrant.

Not that the first opening during a rainy day was perfect. Like many other new structures, the Astrodome had some leaks in the roof. But in keeping with the Texan size of everything, it didn't have just a leak or two. The dome had seventy-three holes that had to be located and sealed off. They were, and now the stadium is dry and weatherproof in the worst storms.

Under the great dome have been bloodless bullfights, rodeos, vacation and travel shows, musical shows, and polo games. There have been auto races and thrill shows, championship boxing, soccer games, political and other types of conventions, and basketball games. The Astrodome has been the scene of track and field meets, lacrosse games, gymnastics, and many other events. In most cases the top stars in each field are involved, and in

The Astrodome has become one of the most important arenas in the sports and entertainment world.
(Houston Sports Association)

many cases the events have taken place for the first time ever under a roof.

It is the home of the Houston Astros baseball team, the Houston Oilers football team, the University of Houston Cougars, the annual Astro Blue Bonnet Bowl football classic, the famous Houston Livestock Show and Rodeo, and many other permanent attractions.

The Astrodome is now part of a complex called Astrodomain. It includes a family amusement park called Astroworld, as well as Astrohall, Astroarena, and four great hotels. One of these is the Astroworld Hotel, and in keeping with the size of the complex, one suite in the hotel is very special. Furnished with valuable antiques, featuring a wine cellar with the finest wines, and a two-level mini nightclub *in the rooms,* the suite rents for $2,500 *per day. The Guinness Book of World Records* called it "the most expensive hotel suite in the world."

The entire Astrodomain complex of attractions, which covers 260 acres of what was once wasted marshland, now brings in more than 100 million dollars to the area each year.

"And, The Best Is Yet To Come," says the slogan of Astrodomain. From a few scratchings on an envelope made by a dynamic man on a visit to Rome, the area has become one of the most important in the sports and entertainment world.

THE MOON LANDING

During the warm summer Sunday evenings of 1969, there was a popular adventure program on television. It concerned a group of superscientists and their difficult missions for the United States. The amazing successes of these agents resulted from daring, cool calculation, and a marvelous blending of scientific equipment, computer logic, and split-second timing. The program was called *Mission Impossible*.

On one particular Sunday evening, though, the exciting program was forgotten because of another, somewhat similar program. Nearly every television receiver in the country was tuned to the new program. Nearly every television channel in the world, for the first time in history, was covering the same event. Across the United States and across the world people gathered to watch the astounding thing happening on television. It was a *new* mission impossible.

To this day some people feel that the program was just an adventure show, that what they were seeing was not real, that it was a pretense shown only to excite the public. These people still believe that the mission they were seeing on television was truly impossible and never really happened.

It did appear almost impossible, but it *was* happening.

It all began near dawn on a foggy, damp morning years before, on May 5, 1961. Four mournful blasts of a foghorn across the flat swampland of mid-coastal Florida announced that all was in readiness for man's first adventure in space. A brave navy officer, Alan B. Shepard, was being strapped into the nose of a ballistic missile for a wild ride downrange. It was exactly like riding in a *bullet*, but the flight was successful.

The adventure that began that day hasn't ended yet, but it has paused along the way for other suborbital ballistic missile rides; short orbital flights around the earth, beginning with John

Moon-bound, the giant cannon, the Saturn V rocket, is launched from Pad A, Launch Complex 39, Kennedy Space Center, on July 16, 1969. Far up in the nose, in the command module that will be the only part of the great vehicle to return, are the astronauts. (NASA)

Glenn's dangerous first attempt; and three-man orbital rides lasting for days and then weeks. Then came deep space flights where men dropped into orbit around the moon.

The adventure paused in sadness at other places not in space. Americans wept as on January 31, 1967, astronaut Edward H. White III returned to rest on a shaded hillside at West Point, New York. Astronauts Virgil "Gus" Grissom and Roger B. Chaffee were brought to Arlington National Cemetery, where many other fallen heroes are buried.

Summing up America's pride and grief, the West Point Cadet Chorus sang from their alma mater: "And when our work is done, our course on earth is run, may it be said, 'Well Done; be thou at peace,' " over the grave of Ed White. The three had died during America's race into space.

All of this, the flights into space and the deadly risks, was aimed at one final triumph. Man wanted to land on another body in space; he wanted to walk upon and scoop up the soil of our nearest neighbor, the moon. It has long been believed by some that eventually man will move out from Earth to colonize other planets. The moon must be the first step, the jumping-off point to deep space.

"The moon is an old target of the dreams of mankind," said Dr. Kurt A. Debus, director of the Kennedy Space Center. "To go to the moon is symbolic of man's leaving the earth, the opening of a vast new frontier."

"The Yankees," said science-fiction writer Jules Verne years ago, "will be the first to fly to the moon because they are engineers by birthright, with a special genius for gunnery." Verne had envisioned men crawling inside a bullet and being "shot" all the way to the moon. He wasn't far wrong.

The cannon was to be the stupendous Saturn V rocket, over 300 feet tall and with 7 million pounds of thrust. The bullet was to be the Apollo spacecraft attached far up on the nose, 364 feet in the air.

Riding in the bullet would be three men, selected almost at random much earlier in the program before scientists knew exactly which Apollo flight would be the one to land on the moon. There would be several orbiting Apollo flights first.

Finally Apollo 11 was selected to make the landing. The spacecraft commander would be civilian Neil Armstrong, veteran of Gemini 8, an earlier two-man orbital flight. Armstrong, a flying buff since childhood, was from Wapakoneta, a tiny village south of Lima, Ohio. He had been a test pilot on the X-15 program. He was quiet, calm, all business and no nonsense. He spoke seldom, but when he did, others listened. He was thirty-eight years old.

The pilot of the lunar excursion module, the part of the spacecraft that would detach and drop to the surface of the moon, would be Edwin "Buzz" Aldrin. Aldrin was known as a perfectionist by the other astronauts, a man with infinite knowledge about space and spacecraft. He was the space walker from Gemini 12, the last of the two-man flights orbiting the earth much earlier. He was thirty-nine years old.

Piloting the command service module that would remain in orbit around the moon would be thirty-eight-year-old Michael Collins, a pleasant man with a ready smile, who was very popular among the astronauts. Another X-15 pilot, Collins was once grounded due to a bone spur in his neck. There were two ways to handle the matter. One was more certain to cure the problem once and for all, but it was more dangerous than the other. Typically, Collins chose the way which could get him back into space, even if it was the more dangerous. The surgery was successful, and he was selected for the moon-landing flight.

The mission looked logical on paper. The Saturn

would hurtle the Apollo spacecraft into Earth orbit, then on into space toward the moon. Early in the seventy-hour journey, the command service module (CSM) would undock from the Saturn, turn end to end in space, and pluck the lunar excursion module (LEM) from its housing in the remaining third stage.

After any necessary mid-course correction maneuvers, the docked CSM and LEM would finally fall into a long sweeping curve around the back side of the moon.

Then they would be slowed with an engine burn to fall into lunar orbit. Armstrong and Aldrin would crawl from the CSM into the LEM. The LEM would separate from the CSM; then braking rockets would be fired to cause the mooncraft to fall to the surface of the moon, where a thrust from the engine would cushion it for a soft landing. Meanwhile, Collins would manage the CSM in lunar orbit.

After lunar exploration, the top half of the LEM would rocket away from the bottom half, carrying the two explorers back to the orbiting CSM. The two spacecraft would dock above the moon, the men would gather in the CSM, and the LEM, having done its job, would be cast loose to fall back to the surface of the moon. One of the sensitive instruments left behind would probably record the shock waves of the crash.

The command service module would burn its engine, carrying the spacecraft out of lunar orbit and toward the earth. After the long journey back, the service module would be discarded in space, and the command module would be turned so that the heat shield would be at the front of the craft to protect the astronauts from heat. Finally, the spacecraft would burn its way back into the atmosphere of Earth, a tiny fraction of the huge Saturn that had left Earth a few days before.

Parachutes would lower the command module to

the ocean, and the great adventure would be over.

That's how it was all supposed to work.

That's how it *did* work.

On that warm Sunday evening in 1969, most of the world had gathered to watch what was happening on television. It was estimated that two out of every three Americans were watching. The people of Poland reported "good reception." The Russians were watching and sent a string of cablegrams to the American space center. The British watched, broadcasting the sound of alarm clocks to wake any dozing viewers so they wouldn't miss the great event. The French watched and cheered. Great "moon watch" gatherings were held everywhere. Telescopes normally pointed toward deep space were focused on one small spot on the moon, on the Sea of Tranquillity.

On that spot sat a spacecraft with men inside. Outside the spacecraft was a television camera. It was attached to the side and pointed at a ladder and a footpad of the spacecraft. Soon it would be switched on, and the world 250,000 miles away was waiting.

The mission had gone according to plan with a thundering lift-off of stunning power and fury. The seventy-hour trip to the moon had been smooth. Armstrong and Aldrin had crawled through the tunnel into the LEM and undocked. While Collins piloted the CSM above, the two had started their descent to the surface of the moon.

Below them lay a panorama of silent rocks, rills, and craters—almost white-lighted in the sun's glare where it could reach, and dark as blackness where shadows lay. It was a forbidding, dangerous scene, and yet calmly beautiful in a stark way.

The time had come to burn the LEM engine and drop out of moon orbit. Man's greatest adventure, his own *real* mission impossible, was ready for the final actions. Anything could happen, for what these astronauts were doing had never been done before. Nobody really knew what might happen next throughout the mission, for it was being done untested. For all they knew, they would sink into the surface of the moon, never to be seen again (some respected scientists insisted this would happen the instant the LEM touched down).

Down came the spidery vessel that would burn to a cinder in the heavier atmosphere of Earth. In the airless void of the moon, scientists were sure it could descend safely. And it was coming down as planned.

The voices of Armstrong and Aldrin crackled from receivers on Earth as the LEM came closer and closer to the moon's surface. Their voices were clipped, their language precise. The time for speculation and planning was long past. The chips were down, and the dice had been cast. There was no way they could turn back.

In space language they reported their progress to Earth and to tape recorders . . . in case something did go wrong.

"Balance couple on . . ." snapped the voice of LEM pilot Aldrin from above the lunar surface. "TA throttle . . . minimum throttle . . . Auto . . . CDR . . ." he continued to report calmly.

Down came the LEM and the two men, closer and closer to the moon.

The operations control room in Mission Control Center in Houston during the moon landing (NASA)

". . . 2000 feet . . . into the AGS . . . 37 degrees . . . 750 coming down to 23 . . . 700 feet . . . 21 down, 33 degrees. . . ." The voice of the LEM pilot droned on, each report indicating something specific to mission controllers on the ground. There, dozens of scientists were hunched tensely over instruments, video lights flickering off their worried faces. This, they knew, was it.

". . . 540 down to 30 . . . 15 . . . .400 feet down to 9 . . . coming down nicely . . . 200 feet . . . 75 feet . . . looking good . . . 60-second lights on . . . down 2 1/2 . . . forward . . . kicking up some dust . . . big shadow . . . 4 forward . . . 4 forward . . . drifting to the right a little . . . down 1/2 . . . 30 seconds. . . ."

Then came the voice of Armstrong, elated but still matter-of-fact. "Contact light! OK, engine stop. . . . ACA at a detent. . . . Modes control both auto. . . . Descent engine command override off. . . . Engine arm off!

"Houston, Tranquillity Base here. The *Eagle* has *landed!*"

Man was on the moon! Armstrong had used the word *base* to indicate they were safely down. The *Eagle* had become the first human habitation on any body off the earth.

Nobody on Earth had yet *seen* the achievement, but most believed that it was really happening. And soon they would see.

Meanwhile, from the earth, a mission controller who had witnessed many astounding flights before and learned to report calmly about them, lost his composure in a happy outburst. "Roger, Tranquillity," he answered, "we copy you *on the ground!* You've got a bunch of guys about to turn *blue!* We're *breathing* again! Thanks a lot!"

Television receivers throughout the world waited quietly, for more was to come. Television announcers, professional talkers who could normally speak about anything without pause, were strangely silent. What was happening spoke for itself. The awesome mission continued as dark television screens waited.

More information crackled down. The pilots of the LEM had been required, at the last instant, to take over and manually fly the craft to the landing because of huge rocks in the primary area. "Houston," reported Armstrong, "that seemed like a very long final phase. The auto target was taking us right into a football-field-size crater with a lot of boulders. It required us flying manually over the rock field to find a real good area."

First the astronauts described the forbidding terrain outside their ship; then, unable to wait through a programmed rest period, they began to don their $300,000 spacesuits. The suits would protect them in the airless void outside, where the temperature was a scalding 250 degrees above zero in the sun and 250 degrees *below* zero in a patch of shade just a step away. Air serves as a buffer and temperature moderator on Earth, but there was no air on the moon.

Finally, the moment arrived. The cabin was depressurized to match the outside, and the main hatch was opened. Slowly, with great care so that his protective suit would not catch on any projection and rip (that would mean instant death), Armstrong edged out onto the high "front porch"

Astronaut Edwin Aldrin, the LEM pilot, steps down to the surface of the moon. The photo was taken by Armstrong during the crew's extravehicular activity. (NASA)

platform of the LEM. Aldrin assisted him from inside as his foot reached for the ladder leading to the moon ground. His hand reached for the D ring that would, when pulled, activate the exterior TV camera.

Armstrong pulled the ring. On Earth, a beautiful white and blue basketball in the sky over the lonely LEM, television screens flickered. Light blinked and faded; then slowly a dim and hazy image came on the screens. It was the astronaut Armstrong. He could be *seen!* Upside down at first until technicians corrected the signal coming from the moon, then right side up, he carefully reached with his foot for the rungs of the ladder. Slowly he descended. With deliberation he dropped to the footpad of the spacecraft; then he prepared to step off onto the ground.

He made the first step of a human onto a body in space.

"That's one small step for a man, one giant leap

Moving about in a strange, kangaroolike gait, the astronauts set out scientific instruments that will be left behind. (NASA)

for mankind." His voice came crackling over the radios and television receivers tuned by the millions to the adventure. Then Aldrin appeared on the ladder and came down to join his commander.

The great mission had succeeded, no matter what happened next. Man was on the moon. Soon both astronauts were moving about confidently in a strange, kangaroolike gait (due to the gravity of the moon which is one-sixth that of Earth), placing scientific instruments that would be left behind to collect data, and gathering about seventy pounds of rock and dust to bring back home.

The astronauts planted an American flag and uncovered a silver plaque attached to the part of the LEM that would remain on the moon. It said "Here men from planet Earth first set foot upon the moon, July, 1969. We came in peace for all mankind." Following were the signatures of the three astronauts and the president of the United States.

Neil Armstrong beside the American flag the astronauts left on the lunar surface. The flag is wrinkled, not waving as it appears to be, because there is no air on the moon. Aldrin shot the photo. (NASA)

Other items were left behind. There was a small disk upon which messages from the leaders of seventy-six nations on Earth had been etched. Left behind were mementos from three other astronauts, Gus Grissom, Ed White, and Roger Chaffee. The astronauts on the moon also left behind, as a tribute to Russia's spacemen, medals from two Soviet cosmonauts who had died in space exploration.

All of this while the world watched, for the astronauts had placed a television camera some distance away from the LEM to survey the entire scene. It was one of the most astounding sights ever on television, far more exciting than the fictional *Mission Impossible* that had been canceled for the moon "special."

An earthquake recorder, known from then on as a moonquake recorder, was put into place to record and send to Earth information on lunar seismic activity. A laser beam reflector was aimed at Earth

Sharply etched into the moon soil is the print of an astronaut's boot. This is the soil some scientists insisted would not support the weight of a spacecraft or even a man. (NASA)

for further experiments in the coming months. Rock sample boxes were hoisted aboard the upper section of the LEM. Film from cameras followed, with Aldrin hauling them up and in with an old-fashioned block and tackle. The mission was drawing to an end.

So a new tension began. Would the "impossible" mission continue as perfect as it had been? Would the spacemen succeed in rejoining their waiting comrade far above, and would the three leave moon orbit and return to Earth without problems? Everything had to continue working just right, for there was *no* margin for error.

At 1:00 P.M. on Monday, after a rest period, the controls were activated, and the lift-off button was pushed. With a swooshing roar, the upper section of the LEM blasted away from the lower section, and the spacemen were on their way to rendezvous with Collins in the orbiting CSM.

At 5:35 P.M. the two ships docked, the moonmen

gathered in the CSM, and the LEM upper section was undocked and discarded. At 12:56 A.M. on Tuesday, while they were behind the moon and out of contact with stations on Earth, the astronauts activated instruments and engines. The trajectory was perfect. They were on their way home.

Ed Aldrin radioed a prayer of thanks as the craft streaked through cold space, and Collins spoke a message of peace on earth. Armstrong passed on the gratitude of the crew to all Americans, for they had made the mission a reality. Then (smuggled aboard during preparations for the mission days before) came the tape-recorded sound of a mournful freight train's steam whistle heard from across a prairie. The astronauts were highballing home.

Once again the world watched as the first three moonmen in history prepared for the final dangerous part of their mission, the splashdown. Dawn was just breaking in an area of the Pacific Ocean 400 miles south of Johnston Island. The aircraft carrier U.S.S. *Hornet* waited. Then came the first visual contact with the spacecraft from an aircraft flying high over the area. The pilot had seen a fireball as material from the heat shield burned away leaving a long trail of fire.

Inside the small capsule—all that remained after they had undocked from the service module not long before—the astronauts were safe and secure.

Down through the exact entry corridor they came to splash into the ocean only eleven miles off the bow of the recovery ship *Hornet*. Flotation bags inflated to steady the command module in the water, while inside, the astronauts were donning "biological isolation" garmets. Outside, men in the same type of suits were washing down the spacecraft with a germ-killing solution. Scientists at that time were still not certain whether dangerous bacteria could be carried from the moon to Earth, so precautions were being taken. The astronauts were quarantined for twenty-one days after the mission in a germ-free laboratory as a further precaution.

But the moon had been conquered, and finally, the Apollo 11 crew was released to an admiring world. Everywhere they traveled, people cheered. Great parades and banquets were held, always with world leaders in attendance. More people would visit the moon in years to come, but these three had been the first. They had blazed the trail.

There are those who *still* believe that it has not been, and cannot be, done.

The reentry was perfect. The three astronauts and a frogman are outside the command module in a rubber life raft, awaiting pickup by a helicopter. All are wearing sealed garments in case they had become contaminated while on the moon. (NASA)

Suggested Readings

Bloemker, Al. *500 Miles to Go: The Story of the Indianapolis Speedway*. New York: Coward, McCann & Geoghegan, 1961.

Casey, Robert J., and Borglum, Mary. *Give the Man Room*. Indianapolis: The Bobbs-Merrill Co., Inc., 1952.

Fite, Gilbert C. *Mount Rushmore*. Norman, Oklahoma: University of Oklahoma Press, 1964.

James, Theodore, Jr. *The Empire State Building*. New York: Harper & Row, Publishers, Inc., 1975.

McCullough, David. *The Great Bridge*. New York: Simon and Schuster, Inc., 1972.

Olney, Ross R. *Americans in Space: A History of Manned Space Travel*. New York: Thomas Nelson, Inc., 1966.

——— *Daredevils of the Speedway*. New York: Grosset & Dunlap, Inc., 1966.

Steinman, David B. *The Builders of the Bridge: The Story of John Roebling and His Son*. New York: Harcourt Brace Jovanovich, Inc., 1950.

Zeitner, June C., and Borglum, Lincoln. *Borglum's Unfinished Dream—Mount Rushmore*. Aberdeen, South Dakota: North Plains Press, 1976.

Index

Italic page numbers refer to illustrations.